Protein localization by fluorescence microscopy

The Practical Approach Series

SERIES EDITOR

B. D. HAMES
Department of Biochemistry and Molecular Biology
University of Leeds, Leeds LS2 9JT, UK

See also the Practical Approach web site at **http://www.oup.co.uk/PAS**

★ **indicates new and forthcoming titles**

Affinity Chromatography
Affinity Separations
Anaerobic Microbiology
Animal Cell Culture
  (2nd edition)
Animal Virus Pathogenesis
Antibodies I and II
Antibody Engineering
Antisense Technology
★ Apoptosis
Applied Microbial Physiology
Basic Cell Culture
Behavioural Neuroscience
Bioenergetics
Biological Data Analysis
Biomechanics – Materials
Biomechanics – Structures and
  Systems
Biosensors
★ Caenorhabditis Elegans
Carbohydrate Analysis
  (2nd edition)
Cell-Cell Interactions
The Cell Cycle

Cell Growth and Apoptosis
★ Cell Growth, Differentiation
  and Senescence
★ Cell Separation
Cellular Calcium
Cellular Interactions in
  Development
Cellular Neurobiology
Chromatin
★ Chromosome Structural
  Analysis
Clinical Immunology
Complement
★ Crystallization of Nucleic
  Acids and Proteins
  (2nd edition)
Cytokines (2nd edition)
The Cytoskeleton
Diagnostic Molecular
  Pathology I and II
DNA and Protein Sequence
  Analysis
DNA Cloning 1: Core
  Techniques (2nd edition)
DNA Cloning 2: Expression
  Systems (2nd edition)

Protein localization by fluorescence microscopy

A Practical Approach

Edited by

VICTORIA J. ALLAN
School of Biological Sciences,
University of Manchester,
2.205 Stopford Building, Oxford Road,
Manchester M13 9PT

OXFORD

UNIVERSITY PRESS

OXFORD

UNIVERSITY PRESS

Great Clarendon Street, Oxford OX2 6DP

Oxford University Press is a department of the University of Oxford
and furthers the University's aim of excellence in research, scholarship,
and education by publishing worldwide in

Oxford New York

Athens Auckland Bangkok Bogotá Buenos Aires Calcutta
Cape Town Chennai Dar es Salaam Delhi Florence Hong Kong Istanbul
Karachi Kuala Lumpur Madrid Melbourne Mexico City Mumbai
Nairobi Paris São Paulo Singapore Taipei Tokyo Toronto Warsaw

and associated companies in Berlin Ibadan

Oxford is a registered trade mark of Oxford University Press

Published in the United States
by Oxford University Press Inc., New York

A catalogue record for this book is available from the British Library

Library of Congress Cataloging in Publication Data

Protein localization by fluorescent microscopy : a practical approach
/ edited by Victoria J. Allan.
p. cm. — (The practical approach series ; 218)
Includes bibliographical references and index.
1. Proteins—Analysis. 2. Fluorescence microscopy. 3. Green
fluorescent protein. 4. Immunofluorescence. I. Allan, Victoria J.
II. Series.
QP551.P6965 2000 572'.636—dc21 99–37444

ISBN 0-19-963741-5 (Hbk)
0-19-963740-7 (Pbk)

Typeset by Footnote Graphics, Warminster, Wilts
Printed in Great Britain by Information Press, Ltd,
Eynsham, Oxon.

*In memory of Thomas Kreis—master of fluorescence microscopy—who died in
the Swissair flight 111 crash on 3 September 1998*

Preface

For a variety of reasons, fluorescence light microscopy is undergoing something of a renaissance. For instance, there is an ever-expanding list of genes that have been sequenced—often through the genome sequencing projects—but which are of completely unknown function. A key requirement of 'functional genomics' is the ability to determine the location of such gene products within the cell, as this may give vital clues about their roles *in vivo*. This can be achieved remarkably simply, and without the need for producing antibodies, by generating 'tagged' proteins which are then expressed in the cells of interest and localized via the tag. In more and more cases, the tag used is the green fluorescent protein (GFP), which then also provides a means for localizing the gene product in living, as well as fixed, cells.

This apparent ease of localization can be misleading, however, since there are plenty of possible artefacts: one of the aims of this book is to provide not only the practical information needed to perform such experiments, but also clear descriptions of the caveats and controls for these approaches in a number of different systems, including vertebrate cultured cells (Chapter 5), *Drosophila* (Chapter 6), plants (Chapter 7) and yeasts (Chapter 8). The range of modified GFPs, and their use, is also outlined.

Before the advent of GFP, one way in which protein localization (and sometimes function) could be investigated in living cells was by the micro-injection of fluorescently labelled antibodies to the protein of interest or, indeed, the fluorescently labelled protein itself. Although it requires a reasonable amount of specialized equipment and expertise, this approach will continue to be of importance, as it provides a means of introducing a GFP-labelled protein (via the micro-injection of DNA) together with an appropriate fluorescently labelled antibodies. These methods are outlined in Chapters 5 and 6.

The use of GFP for localizing proteins must necessarily go hand-in-hand with conventional immunofluorescence, because once the distribution of the novel protein has been observed, it is then crucial to be able to interpret the staining pattern. For example, if the protein is localized to filamentous structures, then which type of filaments are they? On the other hand, if it is membrane-associated, which organelle is involved? To analyse the localization requires expertise in using antibodies to label different sub-cellular structures in the organism of interest. Again, whilst this is straightforward in some cases, there are many ways in which problems can arise: perhaps the antibodies will not recognize their epitopes under the fixation conditions used, or maybe the sub-cellular structure to be studied is not well preserved. Moreover, since a thorough study of protein localization will almost always use multiple antibodies, it is unfortunately common for each antibody to need a specific set of fixation and permeabilization conditions. In addition, if the protein is nuclear, it may be of interest to localize it along with specific DNA domains identified by fluorescence *in situ* hybridization. We therefore provide a detailed description of how proteins and DNA can be localized simultaneously (Chapter 3).

All of the problems described above are exacerbated when one is attempting to localize the protein of interest in its native state using a newly made antibody, rather than by expression of a tagged version of the protein, and a painstaking investigation is sometimes needed to determine the conditions in which the antibody will recognize its epitope. I would hazard a guess that many antibodies have been dismissed prematurely as 'not working for immunofluorescence', simply because the appropriate conditions (or cell type) were not tested. Having established conditions where staining is obtained, all that remains is to determine whether the localization is real! Because of the complexity involved in performing and interpreting such studies, we have included, alongside the basic immunofluorescence methods, an outline of the tricks—and the pitfalls—of immunofluorescence of whole vertebrate cells (Chapter 1); *Drosophila* (Chapter 6) and yeasts (Chapter 8). In addition, Chapter 2 describes the use of semi-thin sections for fluorescence microscopy—an under-used technique that has great potential for tissues and thicker cells, such as polarized epithelia. A list of commercial sources of antibodies to a whole range of sub-cellular structures is provided in Appendix 2.

Of course, for fluorescence microscopy the selection of appropriate equipment is vital. Anyone who is new to the field (or has never had to buy such equipment before) is faced with a bewildering array of possibilities, at hugely varying cost, and it is not a good idea to rely solely on the information provided by sales representatives! If a range of equipment is already available in your Department, then the decision may simply be whether to use a confocal or a conventional fluorescence microscope. While such equipment (if properly maintained) is suitable for imaging fixed specimens, what if you want to study the protein in living cells, or in thick specimens? What filter sets are there, and what combination of fluorescent dyes can be used for double or triple labelling? To help the reader navigate through the maze of choices, we have included information on the range of microscopy hardware and its applications (Chapters 4–6), although the reader should look elsewhere for detailed instructions on the nitty-gritty of using microscopes.

In order to make the most of the more sophisticated approaches—such as confocal or multiphoton microscopy, digital deconvolution or multi-wavelength 4D microscopy—it is also important to understand a number of the underlying principles of microscopy and optics. It should be emphasized that such knowledge will also help any microscopist to make the best of their specimens and equipment. This background is provided in Chapters 4 and 6.

While it is clear that some direct assistance with the microscopy will be needed, this volume aims to provide all the necessary information to enable someone new to microscopy to determine the localization of their chosen protein successfully. In addition, we feel confident that the wide range of techniques covered, including live cell work—and the inclusion of a number of 'tricks of the trade'—will mean that there is also plenty here for the seasoned light microscopist.

July 1999   V. J. Allan

Contents

3. Simultaneous *in situ* detection of DNA and proteins   51

Klaus Ersfeld and Elisa M. Stone

4. Instruments for fluorescence imaging   67

W. B. Amos

Contents

5. Fluorescence microscopy of living vertebrate cells 109

Rainer Pepperkok and David Shima

6. Visualizing fluorescence in *Drosophila*— optimal detection in thick specimens   133

Ilan Davis

Contents

Appendices

Contributors

VICTORIA J. ALLAN
School of Biological Sciences, 2.205 Stopford Building, University of Manchester, Oxford Road, Manchester M13 9PT, UK.

W. B. AMOS
MRC Laboratory of Molecular Biology, Hills Road, Cambridge CB2 2QH, UK.

KATHRYN R. AYSCOUGH
Department of Biochemistry, MSI, WTB Complex, Dow Street, Dundee DD1 5EH, UK.

PETRA BOEVINK
Department of Virology, Scottish Crop Research Institute, Invergowrie, Dundee DD2 5DA, UK.

ILAN DAVIS
Institute of Cell and Molecular Biology, King's Buildings, Edinburgh University, Mayfield Road, Edinburgh EH9 3JR, UK.

KLAUS ERSFELD
School of Biological Sciences, 2.205 Stopford Building, University of Manchester, Oxford Road, Manchester M13 9PT, UK.

IAIN M. HAGAN
School of Biological Sciences, 2.205 Stopford Building, University of Manchester, Oxford Road, Manchester M13 9PT, UK.

CHRIS HAWES
School of Biological and Molecular Sciences, Oxford Brookes University, Oxford OX3 0BP, UK.

IAN MOORE
Department of Plant Sciences, Oxford University, South Parks Road, Oxford OX1 3 RB, UK.

ALISON J. NORTH
School of Biological Sciences, 3.239 Stopford Building, University of Manchester, Oxford Road, Manchester M13 9PT, UK.

RAINER PEPPERKOK
ALMF/Cell Biology/Biophysics Programme, EMBL, Heidelberg, 69117 Heidelberg, Germany.

Contributors

DAVID SHIMA
Cell Biology 624, Imperial Cancer Research Fund, PO Box 123, 44 Lincoln's Inn Fields, London WC2A 3PX, UK.

J. VICTOR SMALL
Institute of Molecular Biology, Austrian Academy of Sciences, Billrothstrasse 11, A-5020 Salzburg, Austria.

ELISA M. STONE
Department of Molecular, Cellular and Developmental Biology, University of Colorado, Boulder, CO 80309–0347, USA.

Abbreviations

3D	three-dimensional
AOD	acousto-optical deflector
APD	avalanche photodiode
BAC	bacterial artificial chromosome
BFP	blue fluorescent protein
BSA	bovine serum albumin
CaMV	cauliflower mosaic virus
CCD	charge-coupled device
CFP	cyan fluorescent protein
CW	continuous wave
DABCO	1,4-diazobicyclo(2,2,2)-octane
DAPI	4′,6′ diamidino 2-phenylindole dihydrochloride
DIC	differential interference contrast
$DiOC_6(3)$	3,3′-dihexyloxacarbocyanine iodide
DMSO	dimethylsulfoxide
DSP	dithiobis(succinimidyl proprionate)
EBFP	enhanced blue fluorescent protein
EGS	ethylene glycol bis(succinimidylsuccinate)
EM	electron microscopy
ER	endoplasmic reticulum
FISH	fluorescent *in situ* hybridization
FITC	fluoroscein isothiocyanate
FRET	fluorescence resonant energy transfer
FWHM	full width–half maximum
GFP	green fluorescent protein
GST	glutathione S-transferase
GUS	β-glucuronidase
HA	haemagglutinin
ICCD	intensified charge-coupled device
MDCK	Madin Darby canine kidney
MOS	metal-oxide silicon
NA	numerical aperture
NGS	normal goat serum
NPG	*n*-propyl gallate
NRK	normal rat kidney
NSOM	near-field scanning optical microscopy
dNTPs	deoxynucleotide triphosphates
OD	optical density
OTF	optical transfer function

PBS	phosphate-buffered saline
PCR	polymerase chain reaction
PEG	polyethylene glycol
PMT	photomultiplier tube
PPD	*para*-phenylene diamine
PSF	point spread function
PVA	polyvinyl alcohol
PVP	polyvinylpyrrolidone
PVX	potato virus X
QE	quantum efficiency
RI	refractive index
SDS	sodium dodecyl sulfate
SIT	silicon-intensified target
SNR	signal-to-noise ratio
SPB	spindle pole body
TBS	Tris-buffered saline
TIRF	total internal reflectance fluorescence
TRITC	tetramethylrhodamine-5-isothiocyanate
ts-O45-G	vesicular stomatitis virus ts-O45-glycoprotein
YAC	yeast artificial chromosome
YFP	yellow fluorescent protein

1

Basic immunofluorescence

VICTORIA J. ALLAN

1. Introduction

Immunofluorescence on fixed samples forms a vital part of most investigations into protein function, since it provides an idea of the whereabouts of a protein within the cell, which in turn can give important clues as to the protein's function. In addition, immunofluorescence can tell us whether this location varies through the cell cycle or development, and whether this is consistent in cells from different tissues or species. All of this information is very valuable, particularly now that genome sequencing data are available. However, in order to obtain meaningful results, and to interpret them correctly, it is important that the methods and limitations of immunofluorescence are understood.

This chapter aims to provide the basic methods needed for performing immunofluorescence on fixed tissue culture cells; subsequent chapters will deal specifically with yeasts, *Drosophila*, trypanosomes and plants, and with the use of tissue sections for immunofluorescence. For other organisms, *Table 1* contains references to detailed methods, although most of the principles will remain the same. Of course, in order to perform immunofluorescence, one needs antibodies; while we will discuss how to use antibodies, and provide a basic list of suppliers of antibodies for immunofluorescence, the reader should refer to one of the excellent laboratory manuals (1) for details of both raising and purifying them.

1.1 Important considerations

It should be stressed that selecting the right fixation and permeabilization conditions will often make the difference between success and failure in immunofluorescence. Finding the best combination may require extensive testing of a variety of methods, cell types and species. Ultimately, however, it may be that a particular antibody will never work for immunofluorescence.

It is also quite common that antibodies to different epitopes on a single protein, or protein complex, may give very divergent results; for instance, although the motor protein cytoplasmic dynein has been implicated by various functional studies to be involved in localizing the Golgi complex to the perinuclear region, most antibodies to a variety of cytoplasmic dynein com-

Table 1. Immunofluorescence methods for different species and cell types

Species	Reference
Caenorhabditis elegans	(2)
Drosophila melanogaster	Chapter 6
Plants	Chapter 7 and (3)
Polarized vertebrate epithelial cells grown on filters	(4)
Saccharomyces cerevisiae	Chapter 8
Schizosaccaromyces pombe	Chapter 8
Sea urchin and other marine organisms	(5–7)
Trypanosoma brucei	Chapter 3
Xenopus laevis eggs and oocytes	(7,8)

ponents label vesicular structures in the cytoplasm (e.g. *Figure 1B*) rather than the Golgi complex (e.g. ref. 9). Recently, though, we have raised antibodies to cytoplasmic dynein heavy and intermediate chains that label the Golgi complex strongly by immunofluorescence, in addition to vesicular structures (*Figure 1A*). The anti-intermediate chain antibody was raised against the N-terminal 96 amino acids of the protein; when the serum was subfractionated into those antibodies reacting with amino acids 1–60 and those reacting with 60–96, the former population gave rise to Golgi apparatus labelling (*Figure 1A*) whereas the latter produced a vesicular staining pattern (*Figure 1B*) typical of previously-described anti-dynein antibodies. It is crucial, therefore, never to assume that one antibody is providing the full picture in terms of the localization of a particular protein. In addition, if an antibody gives distinct staining patterns under differing fixation conditions, then one should be particularly cautious in interpreting such results. In the worst cases, under certain fixation conditions an antibody may give rise to a very clear labelling pattern (e.g. of mitochondria) that is completely artefactual: it is absolutely vital, therefore, to perform all the necessary controls, and to view one's data with healthy scepticism.

Ideally, a number of other methods will be used in combination with immunofluorescence of cultured cells to confirm the distribution of your protein of interest. These might include micro-injection of antibodies into living cells (Chapters 5 and 6), immunofluorescence of sections (Chapter 2), or the production and expression of chimeras of your protein with epitope tags or the green fluorescent protein (Chapters 5–8). Each of these procedures, of course, has associated caveats, which will be discussed.

2. Growing cells on coverslips

Most immunofluorescence on cultured cells is performed on cells grown on glass coverslips. For polarized cells—such as the epithelial cell lines MDCK or

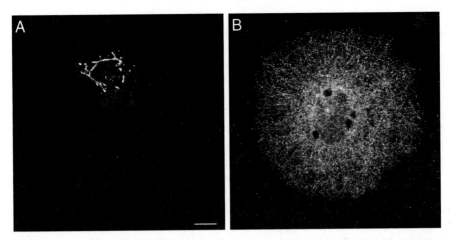

Figure 1. Antibodies to adjacent regions of the same protein may give very different labelling patterns. A sheep polyclonal antibody was raised against the first 96 amino acids of *Xenopus laevis* cytoplasmic dynein intermediate chain. The antibody was then affinity purified against amino acids 1–60 (A) or 60–96 (B) and used for immuno-fluorescence on methanol-fixed (*Protocol 2*) *Xenopus* XTC cells. (A) The 1–60-purified antibody population recognizes the Golgi apparatus (confirmed using a variety of Golgi apparatus markers (not shown)). (B) The 60–96-purified antibody labels small vesicular structures and 'dots'—many of which align with microtubules—but *not* the Golgi apparatus. Images were obtained using a Leica TCSNT confocal scanning microscope equipped with a krypton/argon laser, and are 2D projections of a *z*-series stack of eight images. Scale bar represents 10 μm.

Caco 2—the cells need to be grown under specialized conditions, on permeable filters, in order to become fully polarized. Such methods are described in detail elsewhere (10).

Many cell lines will grow on untreated, sterile coverslips. Using 12 × 12 mm square or 13 mm diameter round coverslips will minimize the amount of antibody needed. The choice of coverslip thickness (or number) is important, as most microscope objectives are optimized for use with glass that is 170 μm thick, which is the average thickness of number 1.5 coverslips. Multi-well slides are available from many suppliers, and many people find them convenient. However, image quality from cells in multi-well slides will not be as good as for cells on coverslips, since the cells will be separated from the coverslip by mounting medium, and therefore will be >170 μm from the upper surface of the coverslip. This disadvantage may be mitigated by using immersion oils of different refractive indices, as described in detail in Chapter 6.

Sterile coverslips should be placed in a tissue culture dish (or in a 6- or 24-well tray, if more convenient) before adding an appropriate dilution of trypsinized and resuspended cells. The dilution will depend on how dense you want your cells on the coverslips, and how soon you will perform the experiment. Clearly, for reproducible results, the cells should be in top condition.

This often means not using cells that have been through too many passages, or that have been left at confluency for too long at any point.

Some cell lines may not grow as well on untreated glass, or may not attach at all without a coating such as poly-L-lysine (*Protocol 1*), or without extra-cellular matrix components, many of which are available commercially. These coatings are likely to result in greater background staining of the coverslip surface, however. Sometimes the preliminary acid-washing step alone (*Protocol 1*) may aid cell attachment. Mitotic cells adhere less well than inter-phase cells, and so may require different conditions in order to make sure they are properly represented within the fixed cell population.

Suspension cells can be attached to coverslips or slides by centrifugation in a cytospin centrifuge, or by incubating the cells on poly-L-lysine-coated cover-slips for 10 min, followed by fixation. Both methods may cause considerable alteration in cell morphology, though, and suspension cells in general are not ideal for immunofluorescence, since they are usually small and round.

Protocol 1.   Coating coverslips with poly-L-lysine

Equipment and reagents

- Coverslips (e.g. 12 × 12 mm, number 1.5)
- Acid/alcohol (1% v/v HCl in 70% ethanol)
- Working solution: poly-L-lysine solution diluted 1/10 in de-ionized water. Store in plastic tubes at 4°C for up to 3 months
- Poly-L-lysine solution (Sigma P 8920), 0.1 % (w/v in water). Store at room temperature
- Coverslip rack and suitable container (e.g. from Agar Scientific Ltd or Molecular Probes)

Method

1. Put the coverslips in the rack and clean by incubating in acid/alcohol for 1 h at room temperature. Remove the solution from the container, and allow container and coverslips to air dry.

2. Incubate coverslips in the poly-L-lysine working solution for 5 min at room temperature.

3. Remove poly-L-lysine solution, drain the coverslips and place indi-vidually on filter paper in a Petri dish. We use the coverslips without further sterilization, and have not had problems with contamination over a few days. They can be UV-sterilized if desired, however, or the whole process can be performed in a laminar flow hood.

---

3. Fixation

An ideal fixation protocol would stabilize all structures within the cell without altering them, prevent loss of soluble proteins and yet allow complete access of antibodies to their epitopes. Needless to say, such a protocol does not exist, and we must compromise in order to balance these conflicting requirements.

Furthermore, a protocol which successfully fixes a given structure in verte-brate cultured cells, such as microtubules, may not be suitable for a mem-branous organelle (*Figure 2*), or for microtubules in other organisms—hence the requirement for other chapters in this volume.

It is also vital to realize that fixation itself can lead to a redistribution of the protein of interest. In a paper that should be required reading for anyone starting to use immunofluorescence, Melan and Sluder (11) introduced a variety of fluorescently labelled soluble proteins into living cells and then compared their distribution before and after a range of common fixation and pre-extraction conditions. The range of changes observed was large, including loss from the cytoplasm, apparent accumulation in the nucleus and association with some sort of cytoskeletal filaments—all induced by fixation conditions. Protein components of membranous organelles or the cytoskeleton are generally *less* liable to such artefacts, but are also preserved to varying extents by different fixation methods (*Table 2, Figure 2*). It is also important to note that even when a single chemical fixation method is used, different membrane proteins may be fixed to a variable degree, so that some membrane proteins remain in place when cells are permeabilized with a non-ionic detergent, whereas others can be almost completely washed away (12).

Two main methods are used for fixation: chemical cross-linking, and protein precipitation via solvent extraction. Each of these methods has their advantages and disadvantages, as described below, and in Chapters 2, 3 and 8. This subject is often covered in great detail in texts on electron microscopy (e.g. refs 13 and 14).

Table 2. Fixation conditions for different structures in cultured vertebrate cells

Cellular structure	Recommended methods	Less suitable methods
Microtubules	Methanol at −20 °C (*Protocol 2; Figure 2*) Methanol/acetone at −20 °C Glutaraldehyde (*Protocol 4*) Formaldehyde + low glutaraldehyde (*Figure 2*) EGS (*Protocol 5*)	Formaldehyde (*Protocol 3; Figure 2*)
Intermediate filaments	Any (*Figure 3*)[a]	
Actin (phalloidin staining)[b]	Formaldehyde	Methanol at −20 °C
Actin (antibody staining)[c]	Methanol at −20 °C	Formaldehyde
Membranous organelles[d]	Formaldehyde Glutaraldehyde	Methanol at −20 °C

[a] Antibody reactivity may be a problem with formaldehyde.
[b] Even though the actin filaments are fixed by methanol to a degree, the phalloidin does not bind to solvent-precipitated filamentous actin. For formaldehyde-fixed, phalloidin-labelled actin filaments, see *Figure 3*.
[c] Antibodies to actin often work better after methanol than formaldehyde.
[d] See *Figures 2* and *4*.

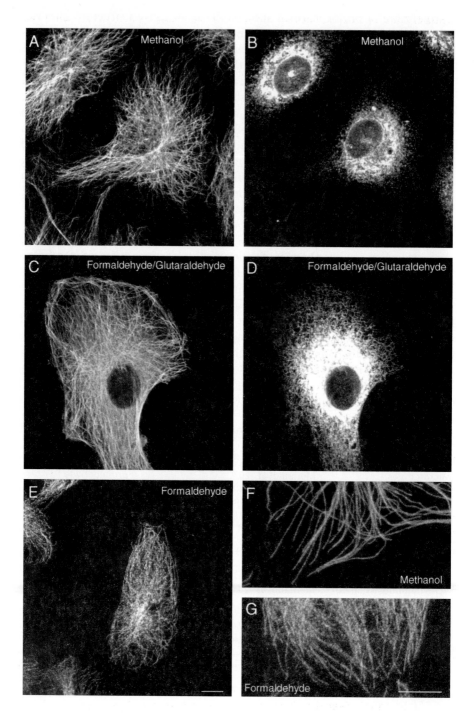

Figure 2. Effect of different fixation conditions on the preservation of microtubules and the endoplasmic reticulum (ER). *Xenopus* cell lines (XTC in panels A–D; XL2 in E–G) were fixed using methanol (panels A, B and F: *Protocol 2*); a mixture of 2% formaldehyde and 0.2% glutaraldehyde (panels C and D: *Protocol 4*) or 3% formaldehyde (panels E and G: *Protocol 3*). Methanol fixation is good for microtubules (A, F), but leads to fragmentation of the delicate tubular structures of the ER (B). Formaldehyde fixes microtubules poorly (E, G), and ER moderately well (not shown). The combination of formaldehyde and glutaraldehyde leads to good fixation of both structures (C and D), but in order to get strong labelling with the anti-ER antibody (mouse monoclonal 1D3, from Dr D. Vaux, University of Oxford), a mixture of 0.1% Triton X-100 and 0.05% SDS was used for permeabilization. Anti-tubulin antibodies used were rat monoclonal YL1/2 (from Dr J. Kilmartin, MRC-LMB, Cambridge: commercially available from SeroTec) for panels A and C, with mouse monoclonal B-5-1-2 (Sigma) used in panels E–G. Images were obtained using a Leica TCSNT confocal scanning microscope equipped with a krypton/argon laser, and are 2D projections of a *z*-series stack of eight images. Panels A, B, E–G were all 1024 × 1024 pixel scans, whereas C and D were 512 × 512. Scale bars represent 10 μm: the bar in E applies to A–E, and the bar in F also applies to panel G.

3.1 Solvent fixation

Fixation in methanol at –20°C is the simplest fixation method for immuno-fluorescence, since no separate permeabilization step is needed, and, because no cross-linkers are used, no quenching of unreacted groups is necessary. The time from removing the coverslip from the culture dish to starting the incubation with primary antibody can therefore be as little as 10 min. It is quite a harsh, denaturing fixation, however, so if your antibody does not give any signal following this method, then do not despair: it might after chemical cross-linking.

Protocol 2. Solvent fixation

Equipment and reagents

- Methanol and acetone (Analar grade)
- PBS (140 mM NaCl; 2.7 mM KCl; 1.5 mM KH₂PO₄; 8.1 mM Na₂HPO₄)
- Fine forceps
- 6- or 24-well culture tray (does not need to be sterile)
- Coverslip rack and glass staining container (e.g. from Agar Scientific Ltd or Molecular Probes)
- Filter paper (e.g. Whatman no. 1)
- Tissue culture cells of choice grown on coverslips in a tissue culture dish

Method

1. Pre-cool the methanol and acetone to –20°C in separate containers (in a spark-proof freezer), each containing an appropriate coverslip rack. This will take several hours, or overnight. If the methanol is not cold enough, fixation will be bad.

2. Gather together the forceps, about 100 ml of PBS in a beaker, and the filter paper. Tear the filter paper discs in half in order to get a more

---

Protocol 2. *Continued*

absorbent edge. Take the dish containing cells on coverslips out of the incubator, and the methanol container from the freezer.

3. Remove a coverslip from the dish using the forceps, dip into the PBS (to wash off excess medium), blot the edge of the coverslip on the filter paper. Wipe excess PBS from the forceps and the back of the coverslip using the torn edge of the filter paper, without allowing the cells to dry out, and then put the coverslip into the rack in the methanol container. Remember which side of the coverslip has the cells on it! Return the container to the freezer to prevent it warming up too quickly (or it can be kept on the bench if you only want to fix a few coverslips).

4. For methanol fixation alone: after 4 min, remove the coverslip and transfer it to PBS in a 6- (or 24-) well tray, keeping the cells uppermost. The coverslip must be deliberately pushed into the PBS, otherwise it will zoom around on the aqueous surface.

5. For methanol/acetone fixation, remove the coverslip from the methanol after 4 min, blot off any excess solvent, then transfer the coverslip into the acetone container (keeping track of which side of the coverslips the cells are on). Fix for a further 4 min at –20 °C, then transfer into PBS as described in step 5. For certain applications, such as monoclonal antibody screening, the coverslips may be air-dried at this point. Such dry coverslips may then be kept at –20 °C.

6. The fixed coverslips may be used directly for antibody incubations (*Protocol 8*), as no further blocking steps are needed. If necessary, the coverslips can be stored in PBS overnight at 4 °C. For longer storage in PBS (not recommended for best immunofluorescence results), add sodium azide to 0.02% (w/v) to prevent bacterial or fungal growth. Alternatively, coverslips may be stored at –20 °C or –80 °C, if care is taken when thawing out (see ref. 15 for details).

3.2 Fixation by chemical cross-linking

3.2.1 Aldehyde cross-linking

(i) Formaldehyde

Formaldehyde is the simplest aldehyde, CH_2O, and is readily soluble in water. Formaldehyde fixative is generally either bought ready made, or is prepared from paraformaldehyde powder (which is a polymer of high molecular weight oligomers of polymethylene glycols), and the two names are often used interchangeably. Formaldehyde fixation is complex, involving the formation of methylene ($—CH_2—$) bridges between a variety of side groups (including amino, amido, guanidino, thiol, phenolic, imidazolyl and indolyl), and is reversible over time (14). It is therefore not a good idea to leave fixed coverslips in PBS for long periods (some people leave them instead in the fixative). We routinely use commercial formaldehyde solution (either standard, which contains a considerable amount of methanol—up to 10%—to prevent re-polymerization of the formaldehyde; or EM grade, which is methanol-free), though for certain tissues or organisms it can be important to use freshly-made formaldehyde (Chapter 2, *Protocol 1*; Chapter 8, *Protocol 4*).

Two methods have been reported that involve the pre-reaction of formaldehyde with lysine/periodate (16) or cyclohexylamine (17) which may improve fixation quality over the use of formaldehyde alone, but these are not commonly used. In addition, the use of a pH shift to facilitate formaldehyde fixation has also been reported (4).

Protocol 3. Formaldehyde fixation

Reagents

- Formaldehyde/paraformaldehyde 37% (w/v) stock solution (**Caution:** hazardous[a]): this may be prepared from paraformaldehyde powder (Chapter 2, *Protocol 1*; Chapter 8, *Protocol 4*), or may be purchased from laboratory suppliers. Methanol-free, EM-grade stocks may be purchased from specialist EM suppliers (e.g. TAAB; Agar Scientific Ltd; EM Sciences)

- Triton X-100
- PBS
- 1 M glycine solution, adjusted to pH 8.5 with 1 M Tris

Method

1. Assemble all equipment and solutions before taking the cells on coverslips out of the incubator.

2. Dilute stock formaldehyde to 3% in PBS and transfer to 6-well trays. One well will hold up to four 12 × 12 mm coverslips. It is convenient to have one well of fixative, with the remaining wells used for PBS washes and subsequent steps. If you are handling a large number of coverslips, then it may be easier to aspirate off solutions between steps instead of transferring coverslips from well to well.

3. Remove coverslips from the culture dish, dip in PBS (in a beaker) to wash off excess culture medium, then place in the formaldehyde well. For this, and all subsequent steps in this protocol, the cells must be uppermost. Fix for 20–30 min at room temperature.

4. Transfer the coverslip into the next well, which contains PBS. Leave for 5 min at room temperature, then transfer again into fresh PBS for 5 min.

5. Block unreacted groups by transferring the coverslips into wells containing 3–5 ml of PBS plus a few drops of 1 M glycine pH 8.5. Incubate for 5 min and then rinse again in PBS. Alternatively, incubate for 5 min in 50 mM NH_4Cl, then rinse in PBS.

Protocol 3. *Continued*

6. Permeabilize cells, if desired, in 0.1% Triton X-100 in PBS for 4 min at room temperature. Some cells or antigens may require up to 15 min. (For other permeabilization conditions, see *Protocols 6* and *7*). Wash twice more with PBS, for 5 min each. The coverslips can then be used for labelling (*Protocol 8*), or may be stored in PBS (*Protocol 2*).

[a] Paraformaldehyde is highly toxic. It is readily absorbed through the skin and is extremely destructive to skin, eyes, mucous membranes and upper respiratory tract. Paraformaldehyde powder and formaldehyde stock solutions should always be handled in a fume hood. Protective clothing and gloves should be worn when handling powder and solutions.

(ii) Glutaraldehyde

Unlike formaldehyde, glutaraldehyde is a dialdehyde ($HOC–CH_2–CH_2–CH_2–COH$), and therefore has two reactive groups for fixation purposes. It fixes much more rapidly and completely than formaldehyde (again via complex chemistry: ref. 14), and so is often the fixative of choice for electron microscopy. However, this efficient cross-linking may lead to problems for immunofluorescence, because the ability of an antibody to gain access to its epitope may be greatly reduced or lost altogether. In addition, glutaraldehyde fixation increases problems with autofluorescence (Chapter 2, section 7). Given the excellent structural preservation obtained with glutaraldehyde, it is still well worth trying a new antibody under these conditions, as long as one is aware of the potential problems. It is best to try a range of concentrations.

Protocol 4. Glutaraldehyde fixation

Reagents

- Glutaraldehyde stock solution (**Caution:** hazardous[a]): available from the suppliers described above
- Triton X-100
- PBS

- Sodium borohydride (**Caution:** hazardous[b]): make up 0.5 mg/ml sodium borohydride in PBS immediately before use. Do not store the solution in a tightly closed container.

Method

The basic procedure for handling cells on coverslips is the same as that described in *Protocols 2* and *3*.

1. Fix the cells for 10–30 min at room temperature in 0.5–2.0% glutaraldehyde, diluted from stock into PBS. The concentration required may vary, and is determined empirically for each antibody.

2. Wash twice for 5 min in PBS.

3. Incubate the coverslips three times for 5 min in the sodium borohydride solution at room temperature to block any remaining reactive groups, and to reduce autofluorescence.

4. Rinse twice in PBS and permeabilize as desired (*Protocols 3* and *6*). The cells are then ready for immunolabelling (*Protocol 8*).

---

[a] Glutaraldehyde is toxic. It is readily absorbed through the skin and is irritating or destructive to skin, eyes, mucous membranes and upper respiratory tract. Glutaraldehyde stock solutions should always be handled in a fume hood. Protective clothing, glasses and gloves should be worn.
[b] Sodium borohydride is flammable on contact with water, harmful by ingestion and if inhaled, and irritating to eyes. Protective clothing, glasses and gloves should be worn.

(iii) Mixtures of aldehydes

A combination of 3% formaldehyde with low concentrations of glutaraldehyde (e.g. 0.05–0.2%) has often proved useful, since the low levels of glutaraldehyde help to fix labile structures (such as microtubules) without compromising antigenicity too much. This is shown clearly in *Figure 2*. For this method, follow *Protocol 4*, except include 2–3% formaldehyde during fixation.

(iv) Buffer selection

The most common buffer used for formaldehyde and glutaraldehyde fixation is PBS. For best preservation of microtubules, however, it may be advantageous to use microtubule-stabilizing buffers (e.g. PEM, *Protocol 6*) instead (e.g. refs 11 and 18). Such Pipes-based buffers are widely used for fixation of EM samples (e.g. ref. 14). An important point to bear in mind is that all of these buffers are designed for mammalian cells: for frog cells, for example, $0.7 \times$ PBS or PEM should be used, or frog Ringer (e.g. ref. 19). After fixation and permeabilization of frog cells, undiluted PBS may be used for subsequent steps.

3.2.2 Other cross-linkers

A variety of other chemical cross-linkers have been used successfully for fixation, particularly of dynamic structures such as microtubules. Ethylene glycol bis(succinimidylsuccinate) (EGS; 20,21) and dithiobis(succinimidyl proprionate) (DSP; 20,22) are both bi-functional reagents that react with free amine groups of proteins, primarily the ε-amino group of lysine residues. EGS and DSP not only provide excellent fixation of microtubules (more like glutaraldehyde than PFA) and membranes, but also maintain antigenicity of many glutaraldehyde-sensitive epitopes. They have the major drawback, however, of being poorly water-soluble, and are therefore difficult to use (*Protocol 5*). Nevertheless, in some circumstances, these cross-linkers may provide the only practical way of obtaining first-rate fixation and double-labelling of very different cellular structures simultaneously. Since it helps to keep all solutions at 37°C for this fixation method, *Protocol 5* is *not* suitable for cells which do not grow at this temperature (e.g. frog cells).

Protocol 5. Fixation using bi-functional cross-linking reagents

Reagents
- Ethylene glycol bis(succinimidylsuccinate) (EGS)[a] (Pierce or Sigma)
- Dry dimethylsulfoxide (DMSO) (e.g. Fluka 41648, over molecular sieve)
- PBS
- Incubator or dry block heater at 37 °C
- 100 µl and 1 ml pipettors with tips
- Forceps and filter paper

Method

1. Make up 100 mM EGS (Pierce or Sigma) in dry DMSO on the day of use. If you use DMSO with too high a water content (e.g. a bottle that was opened a while ago), then the EGS will not dissolve. Discard any cloudy solutions. To avoid this problem, we buy DMSO in bottles containing a molecular sieve to remove water.

2. Warm PBS to 37 °C, and put 900 µl per well of a 24-well tray (one well per coverslip to be fixed). Keep the tray in the incubator, or on a dry block heater.

3. Put PBS in a beaker, and tear filter paper in half.

4. Warm the EGS solution to 37 °C. Add 100 µl of EGS to each well containing PBS, and very quickly pipette up and down with the 100 µl pipettor, and then the 1 ml pipettor, to mix.

5. As quickly as possible, remove cells from the incubator, dip briefly in PBS and wipe off excess with filter paper. Place in the EGS dilution and incubate for 10 min at 37 °C. Some crystals of cross-linker may become visible during this time, but this does not necessarily mean the fixation has not worked. The reason for working quickly, and keeping all solutions at 37 °C, is to minimize this crystallization.

6. Proceed as for formaldehyde-fixed coverslips (*Protocol 3*).

[a] DSP has been used in a similar protocol, but with a 50 mM stock in DMSO being diluted to 1 mM final concentration (20,22)

4. Permeabilization

If the protein of interest is exposed on the external face of the plasma membrane, it may be advantageous to perform antibody labelling on unpermeabilized, aldehyde-fixed cells. In order for antibodies to have access to intracellular structures, however, the fixed cells must be permeabilized. It is possible to combine the two approaches by labelling fixed, unpermeabilized cells first with the plasma membrane antibody, washing, and then fixing again briefly. After permeabilization, the second primary antibody can be used, followed by secondary antibodies. It is worth noting that solvent fixation not only fixes cells but also results in lipid extraction, which means that a separate permeabilization step is not required.

For aldehyde- or cross-linker-fixed cells, there are a number of permeabilization methods. The most common uses the non-denaturing detergent Triton X-100 (*Protocol 2*). It is quite common for antibodies not to recognize their epitopes following such extraction, either because the antibody only recognizes denatured protein, or because the extent of cross-linking of components around the protein of interest—particularly if it is part of a multi-protein complex—prevents the antibody being able to reach its epitope. If this is the case, then it is well worth trying a number of different permeabilization methods (*Protocol 6*) which are rather more or less denaturing, as these have sometimes proved successful (e.g. 4,12,23–25). If the fixation has caused a cross-link in the epitope itself, then these tricks are very unlikely to help. Detergents such as digitonin and saponin—which act preferentially on membranes that are rich in cholesterol (26)—can be useful for immunofluorescence, using antibodies to epitopes exposed within the cytoplasm or on the cytoplasmic face of organelles (although not generally for those within the organelle lumen) (12,26). If your antibody of interest does not work following Triton X-100 permeabilization (*Protocol 2*), then it is worth trying this method.

Another point to bear in mind is that the strength of your immunofluorescence signal does not necessarily relate to the amount of protein present in a simple way: for instance, a great increase in immunofluorescence signal has been observed under conditions where most of the protein of interest has been extracted (12).

Protocol 6. Permeabilization methods that improve antigen accessibility

Reagents

- Triton X-100
- Sodium dodecyl sulfate (SDS)
- Methanol at 0–4°C in a chamber containing a coverslip rack
- 6 M guanidine hydrochloride in 50 mM Tris–HCl, pH 7.5
- PBS containing 0.05% (w/v) saponin

Method 1 (24)

1. Permeabilize formaldehyde- or glutaraldehyde-fixed and quenched cells on coverslips in 0.1% (v/v) Triton X-100 plus 0.05% (w/v) SDS, for 4 min in PBS at room temperature.[a]

2. Rinse twice in PBS for 5 min. The cells are now ready for immuno-labelling (*Protocol 8*). An example of this permeabilization method is given in *Figure 2C* and *D*.

Protocol 6. *Continued*

Method 2 (adapted from ref. 25)

1. After quenching, permeabilize formaldehyde- or glutaraldehyde-fixed cells on coverslips using methanol at 4°C or on ice.

2. Dry off excess PBS from the coverslips using filter paper (without allowing the coverslip to dry out), and put into a coverslip rack in the cold methanol for 5 min.

3. Transfer to PBS at room temperature. The cells are now ready for immunolabelling (*Protocol 8*).

Method 3 (adapted from ref 23)

1. Permeabilize formaldehyde- or glutaraldehyde-fixed and quenched cells on coverslips in 0.1% (v/v) Triton X-100 *(Protocol 2)*.

2. Incubate coverslips in guanidine solution (either in multi-well trays, or by incubating on a drop of solution on Parafilm), then rinse three times in PBS and immunolabel (*Protocol 8*).

Method 4 (e.g. ref. 4)

1. Fix cells on coverslips with formaldehyde and quench with glycine (*Protocol 2*).

2. Permeabilize cells for >5 min in PBS/0.05% (w/v) saponin.

3. Perform all antibody dilutions, incubations and washes as described in *Protocol 8*, except use PBS/saponin in place of PBS.

4. Rinse once in PBS without saponin, then mount *(Protocol 10)*.

[a] This will not work with solvent-fixed cells: they will be washed off the coverslip.

4.1 Pre-extraction before fixation

A method that has commonly been used to visualize cytoskeletal components, such as microtubules, involves a brief permeabilization step *before* fixation (*Protocol 7*), which serves to remove any soluble components such as tubulin dimers (e.g. refs 27–29). Appropriate buffers are used to stabilize the assembled form of the component of interest.

Pre-extraction can also be used to reduce cytoplasmic background for studying membrane-associated proteins (30). In this case, 0.05% (w/v) saponin is used in place of Triton X-100, since it preferentially extracts cholesterol (present mainly in the plasma membrane), leaving internal organelle membranes intact. For some proteins it may be necessary to use lower concentrations of saponin.

Protocol 7. Detergent extraction before fixation[a]

Reagents

- Microtubule-stabilizing buffer (PEM): 1 mM $MgCl_2$, 5 mM EGTA, 80 mM K-Pipes, pH 6.8 (GTP can be added to 1 mM)
- Four small beakers of PEM containing 0.5% (v/v) Triton X-100 at room temperature
- 3.7% formaldehyde in PEM at room temperature
- Methanol at –20°C

Method

1. Rinse each coverslip briefly in PBS.

2. Dip sequentially in four beakers containing PEM + 0.5% Triton X-100, for 3 sec each.

3. Fix in 3.7% formaldehyde in PEM at room temperature (essentially as described in *Protocol 3*), or in –20°C methanol (*Protocol 2*).

[a] Based on the methods described in refs 11 and 27.

---

5. Labelling protocols

Once cells have been fixed and permeabilized, and any unreacted fixative quenched, then the coverslips are ready for antibody labelling (*Protocol 8*). I have not included any incubations with protein solutions such as bovine serum albumin (BSA), serum (from the same species that was used to raise the secondary antibody) or gelatin, which are commonly used to help block non-specific interactions of antibodies with the specimen (see Chapter 2, *Protocol 7*; Chapter 8, *Protocol 2*). Many protocols also call for the antibodies to be diluted in such blocking agents. Tween 20 (at 0.2%) may also be used.

Having adopted the practice of omitting blocking agents whilst in Dr Thomas Kreis' lab., I subsequently tested whether such solutions improved results with any of the antibodies to organelles or the cytoskeleton that we commonly use in the laboratory. I found that blocking solutions in general *increased* the background, which was otherwise almost negligible. If your primary and secondary antibodies are good, then such steps are unlikely to be necessary if you are labelling cultured cells (this is not true for tissue sections: Chapter 2). If in doubt, try it for yourself!

Protocol 8. Antibody labelling

Equipment and reagents

- Primary antibodies plus appropriate fluorescently labelled secondary antibodies
- Parafilm
- PBS

Protocol 8. *Continued*

Method

1. Dilute the primary antibodies in PBS, and put a drop of each antibody onto a sheet of Parafilm that has been taped to the bench surface. 25 µl of antibody solution is plenty for a 12 × 12 mm coverslip.

2. Take a coverslip out of PBS using fine forceps, and blot the edge of the coverslip on filter paper. Wipe excess PBS from the forceps and the back of the coverslip using the torn edge of a piece of filter paper, without allowing the cells to dry out. Carefully place the coverslip, cells downwards, onto the drop of antibody.

3. Incubate at room temperature for 20–30 min.

4. Raise the coverslip off the Parafilm by pipetting PBS (using a blue tip) at the edge of the coverslip, which serves to raise the coverslip from the Parafilm surface. Then lift the coverslip using forceps, and place in PBS in a 6-well tray.

5. Incubate in PBS for 5 min, and repeat twice more. This can be achieved either by transferring coverslips from well to well, or by leaving the coverslip in one well and aspirating off the PBS between washes. With the second method, take care to prevent the coverslip drying out.

6. Repeat steps 1–5, but using diluted fluorescently-labelled secondary antibody. Coverslips are then ready for mounting (*Protocol 10*).

5.1 Antibody quality and preparation

A variety of different types of antibody preparation can be used successfully for immunofluorescence. If the antibody is monoclonal, then culture supernatant may be sufficient (often used undiluted, however); alternatively, ascites fluid (if available) or purified immunoglobulin fractions (see ref. 1) may be used. Be aware that different preparations of the same monoclonal (this also applies to commercially available antibodies) may work more or less well: it is important to titre every new batch carefully. One very important point to remember is that many monoclonals react with more than one protein by immunoblotting, and so are also likely to do the same by immunofluorescence. More worryingly, the possibility exists that a protein recognized by blotting (denatured) is not the same protein recognized by the antibody by immunofluorescence (native). Probably the only way to be sure of a protein localization is to use two or more antibodies, which recognize different epitopes (although they may not always give the same labelling pattern: see section 1 and *Figure 1*). If a polyclonal primary antibody is to be used, then the best approach is to prepare affinity-purified antibodies (see ref. 1). This is less important if the antibody gives a very strong signal and is therefore used at a high dilution. Again, it is important to work out empirically the best dilution to use to maximize the specific signal while minimizing non-specific labelling. If precipitates of antibody are seen on the coverslip surface, then centrifuge the diluted antibodies in a microcentrifuge before use.

Fluorescently conjugated secondary antibodies are widely available. They vary in terms of the host animal used (goat, sheep, donkey, etc.), the fluorophore, and—most importantly—in how extensively the antibody preparation has been cross-adsorbed against serum proteins from other species. For instance, if anti-mouse and anti-rabbit secondary antibodies are to be used together, then they should have been cross-adsorbed against rabbit and mouse serum proteins, respectively. By this means it is even possible to obtain anti-mouse secondary antibodies which do not recognize rat IgG, and vice versa. However, antibodies that have been cross-adsorbed against closely related species are not as efficient at recognizing all classes of antibodies, and should only be used when absolutely necessary. The choice of fluorophore will be discussed below.

5.2 Single versus multiple labelling

The simplest immunofluorescence uses one primary and one secondary antibody. The staining pattern obtained can then be compared with the phase contrast or differential interference contrast (DIC) image of the same cells. Since most fluorescence microscopes are equipped with filter sets for observing more than one fluorophore, it becomes possible to view multiple fluorescently-labelled structures. These may include DNA (labelled with 4,6-diamidine-2-phenylindole dihydrochloride (DAPI; *Protocol 9*) or other dyes), actin (labelled with fluorescently conjugated phalloidin: *Figure 3A*), organelle-specific dyes (e.g. LysoTracker; see Appendix 1), or two or more antibodies. There are a large number of different fluorophores available (Appendix 1 and http://www.fluorescence.bio-rad.com), many of which can be purchased ready-conjugated to secondary antibodies. Before buying any reagents, though, check which filter sets are fitted to the microscope, as they will determine which fluorophores can be visualized (see Chapters 4–6). In addition, some of the dyes listed are designed for use in conventional fluorescence microscopy using mercury illumination, while others are optimized for specific laser lines in confocal or multi-photon microscopy (Chapter 4). The most commonly used combinations are DAPI (seen as blue); fluorescein or Cy2 (seen as green); and Texas Red, rhodamine or Cy3 (seen as red). Far-red dyes, such as Cy5, cannot be visualized by eye, but can be seen by cameras or by confocal microscopy, giving the opportunity for quadruple labelling.

For multiple antibody labelling, it is most common to use primary antibodies from different species. In this case, all primary antibodies can be included in the same incubation (*Protocol 8*), as can all secondary antibodies. It is also possible, though more complex, to use two primary antibodies from the same species: see Chapter 2, section 6.5. If antibodies from closely related species are to be used (e.g. rat and mouse: see *Figure 2A–D*), specially puri-fied secondary antibodies are needed (see section 5.1). One further important point to note is that companies sell secondary antibodies raised in a whole variety of animals, with goat and donkey perhaps being most common. If you would like to use a combination of a sheep primary antibody with, say, a rabbit primary antibody, then ensure that the anti-rabbit secondary antibody to be used was *not* raised in goat, as you will almost certainly get cross-reaction of goat–anti-rabbit antibodies with the sheep IgGs! There are additional controls that need to be performed for multiple labelling approaches in order to check that there is no cross-over between fluorescence channels, as discussed in section 6.

5.3 Mounting coverslips for microscopy

There are two general types of mounting media for immunofluorescence: those that harden with time, and those that do not. The former, such as Mowiol or Gelvatol, are based on polyvinyl alcohol (PVA), which is liquid to begin with and hardens over a few hours. The latter are usually glycerol-based, and for long-term storage the coverslip will need to be sealed around the edges using nail varnish. Common recipes are given in *Protocol 10*. See Chapter 2, *Protocol 7*, for an alternative recipe for PVA mountant.

Another important feature of mounting media is the inclusion of reagents that inhibit fading (photobleaching) during fluorescence observation. Photo-bleaching is a particular problem for fluorescein, but is seen with all fluoro-phores to some extent. Commonly used reagents and their use are described in *Protocol 10*. A number of commercial mounting media are available (e.g. from Vector Laboratories, or Molecular Probes), which already contain re-agents to inhibit fading (although their identity is not revealed!). See Chapter 2, section 8, for a further discussion of the different anti-fade reagents. We routinely use Mowiol mountant containing 25 mg/ml 1,4-diazobicylco(2,2,2)-octane (DABCO).

Protocol 9. Preparing Mowiol or Gelvatol mounting medium[a]

Reagents

- Mowiol 4-88 (Calbiochem or Hoechst) or Gelvatol 20-30 (Air Products or Monsanto Chemicals)
- Glycerol
- 0.2 M Tris, pH 8.5

Method

1. Add 2.4 g of PVA (either Mowiol or Gelvatol) to 6 g of glycerol in a 50 ml screw-topped plastic tube. Stir to mix, then add 6 ml H_2O, mix again and leave at room temperature for several hours, mixing occasionally.

---

2. Add 12 ml of 0.2 M Tris, pH 8.5, mix, then heat at 60°C for 10 min. Mix occasionally.

3. Remove any undissolved PVA by centifugation at 5000g for 15 min at room temperature.

4. Aliquot, and store at −20°C. Anti-fade reagents and DAPI may be added once thawed (*Protocol 10*).

[a] Based on methods described in refs 28 and 31.

Protocol 10.   Anti-fade mounting media[a]

Reagents

- Mowiol or Gelvatol stock (see *Protocol 9*) containing 25 mg/ml DABCO[b] (Sigma). This mixture can be stored at 4°C for several weeks without losing activity.
- *or* Mowiol stock (180 μl) gently mixed, just before use, with 20 μl *para*-phenylene diamine[b] (PPD; Sigma) dissolved to 10 mg/ml in water. The PPD stock can be stored as 20 μl aliquots at −20°C.
- For a glycerol-based mounting medium[c] containing PPD, see Chapter 8, *Protocol 1*

- *n*-Propyl gallate[b] (Sigma), 2% (w/v) in glycerol.[c] Adjust pH to 8.0. *n*-propyl gallate can also be added to Mowiol at 2.5 mg/ml, but it may interfere somewhat with the solidification of the PVA.
- 100 ng/ml DAPI can be included in any of the mountants to allow visualization of DNA

Method

1. Label frosted microscope slides using pencil (pen inks are fluorescent, and cause problems if they get into the immersion oil on the microscope objective).

2. Take coverslips after their final PBS wash and dip them briefly in water. Wipe off excess water from the back and edges of the coverslip, and from between the forcep blades, using filter paper. Too much drying will cause the cells to flatten; too little will result in poor anti-fade performance, because it will have been diluted.

3. Place the coverslip (cells downwards) onto a drop of the mounting medium of choice on the microscope slide. For a 12 × 12 mm coverslip, a 6 μl drop of Mowiol mountant is sufficient.

4. Allow the Mowiol to set (a minimum of 1 h, preferably longer or overnight) in the dark at room temperature.

5. If a glycerol-based mountant is used, then seal around the edges of the coverslip using nail varnish.

[a] Based on methods described in refs 4 and 32–34.
[b] DABCO, PPD and *n*-propyl gallate are harmful by inhalation, ingestion or skin absorption. Wear appropriate gloves and safety glasses, and handle stock in a fume hood.
[c] All three anti-fade reagents may be combined in a glycerol-based mounting medium, containing 25 mg/ml DABCO, 2.5 mg/ml *n*-propyl gallate and 1 mg/ml PPD.

6. Controls

In order to be able to interpret an immunofluorescence experiment, a number of controls are needed, as listed below. Some of these should be performed every time, but others may be left out for convenience once the quality of the primary and secondary antibodies has been determined.

(a) Use positive controls to show whether the fixation and permeabilization was successful. Commercially available anti-tubulin antibodies are useful in this context, and we always include them in every experiment. This also shows that the secondary antibody is working well.

(b) Check the specificity of your primary antibody by:

　　(i) pre-adsorbing the antibody against its antigen;

　　(ii) comparing with pre-immune serum, if the primary antibody is poly-clonal; if pre-immune serum is not available, then a general IgG fraction (e.g. from Sigma) may be used;

　　(iii) if possible, using a cell type which you know does not express your protein of interest.

(c) Test for secondary antibody quality by:

　　(i) omitting primary antibody but including secondary antibody (detects non-specific binding); this will also reveal autofluorescence;

　　(ii) using the 'wrong' secondary antibody (e.g. anti-rabbit, with a mouse primary antibody) to check the species cross-reactivity; this is par-ticularly important if you are using mouse and rat monoclonals as primary antibodies, with anti-rat and anti-mouse secondary antibodies.

(d) Check the filter sets on the microscope by observing samples labelled with a single fluorophore in the 'wrong' channels. For instance, look at a rhodamine-labelled sample using the fluorescein filter set: if it has a broad (or long-pass: Chapters 4 and 6) emission filter, then you will see a weak signal in this channel.

(e) Test for autofluorescence by omitting both primary and secondary anti-bodies. This is particularly important if glutaraldehyde fixation has been used. It is very common to see a mitochondrial pattern (stronger or fainter, depending on the concentration of fixative used) under these con-ditions. For more information on autofluorescence, see Chapter 2, section 7. If you have trouble with autofluorescence (particularly if using living cells), then try a few different filter sets to find which gives least auto-fluorescence, and choose fluorophores to match the filter set.

7. Double labelling as a means for localizing your protein within the cell

It is quite common to be working on a protein whose location within the cell is not known. Once you have raised an antibody to the protein (or tagged it with an epitope marker, or GFP; see Chapters 5–8), then you need to identify the structures to which it binds. The major suppliers of antibodies that can be used for this purpose are given in Appendix 2.

7.1 Cytoskeletal structures

If the antibody labels filamentous structures, then the protein of interest may be cytoskeletal (*Figures 2* and *3*). There is a wide range of commercial antibodies to microtubules and intermediate filaments which can be used to test which filament type is involved, and fluorescently labelled phalloidin will reveal whether your protein co-localizes with actin filaments. Points to remember are:

- there are a number of different intermediate filament types, with different morphology;
- many pre-immune sera contain antibodies to intermediate filaments and centrosomes;
- stable microtubules are morphologically distinct from the general population of microtubules that are recognized by most anti-tubulin antibodies;
- fixation is a particular issue for cytoskeletal proteins (*Table 2*).

7.2 Membranous organelles

If your antibody recognizes a non-random dotted, 'blobby' or reticular pattern, then organelle markers should be tested (*Figures 1, 2* and *4*).

7.2.1 Endoplasmic reticulum

The endoplasmic reticulum (ER) is continuous with the nuclear envelope, and so any staining pattern that is reticular, with nuclear envelope also stained, may well be ER-associated (*Figure 2*). Important issues are:

- the ER is difficult to fix properly (particularly with methanol: compare *Figure 2* panels *B* and *D*), and so the extensive reticular structure may be lost;
- few, if any, antibodies to the ER are commercially available;
- different sub-regions of the ER may exist.

In the absence of an antibody, you can get an idea of what the ER looks like in the cells you are using by incubating living cells with lipophilic dyes such as $DiOC_6(3)$ (35), although this dye labels mitochondria more intensely than the ER. This labelling does not survive detergent extraction or methanol fixation.

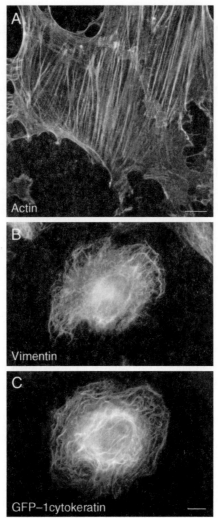

Figure 3. Actin and intermediate filaments in *Xenopus* cells. XTC cells in panel A were fixed with formaldehyde (*Protocol 3*), permeabilized with Triton X-100 and labelled with Texas Red-conjugated phalloidin (Molecular Probes) to reveal filamentous actin. Intermediate filaments in XL2 cells are shown in panels B and C. Cells were transfected with an EGFP–cytokeratin 8 construct, and the EGFP–cytokeratin 8 protein co-assembled with endogenous cytokeratins into cytokeratin filaments (E. Clarke and V. Allan, unpublished data). Cells were then fixed using methanol (*Protocol 2*) and labelled with antibodies to vimentin (clone 14h7, obtained from the Developmental Studies Hybridoma Bank). The vimentin (B) and cytokeratin (C) filament networks are clearly distinct. Panel A was obtained using a Leica TCSNT confocal scanning microscope equipped with a krypton/ argon laser, and is a 2D projection of a *z*-series stack of eight images. Panels B and C were collected using a cooled CCD camera CH250 (Photometrics; 1317 × 1035 pixels) attached to a Leica DM RXA microscope, using a 63× Plan-Apo (NA 1.32) objective. Image acquisition was performed using IPLab Spectrum software on a PowerMac. Scale bars represent 10 μm: panels B and C are at the same magnification.

7.2.2 Golgi apparatus, endosomes and lysosomes

Testing for these organelles can be particularly problematic for a number of reasons:

- there are few commercially available antibodies to organelles in the secretory and endocytic pathways;
- antibodies to the Golgi apparatus, endosomes and lysosomes are often species-specific;
- there may be organelles or organelle sub-compartments that are not yet identified owing to lack of characterized markers;
- some organelles can vary in position and morphology depending on conditions such as intracellular pH or cell cycle, or between different cell types;
- perinuclear labelling is not necessarily Golgi-associated—the intermediate compartment, endosomes and lysosomes can all be found there under certain conditions (*Figure 4A–C*).

It should be emphasized that immunofluorescence can only give you a broad indicaton of the sub-cellular localization of your protein, even when you are comparing the unknown antibody with well characterized organelle markers: immunoelectron microscopy is needed for more detailed analysis (a point of view expressed forcefully in ref. 36).

The Golgi apparatus may be labelled by fluorescent ceramide or brefeldin A derivatives (Molecular Probes: see Chapter 5), but these protocols will not work in fixed, permeabilized cells. Endosomes and lysosomes may be labelled using a variety of fluorescently labelled molecules which are taken up by endocytosis before fixation, or by reagents that accumulate in such acidic organelles. Some of these reagents can be fixed. Molecular Probes sell a wide range of such reagents, and provide a large reference list for their use. Their catalogue and website (http://www.probes.com) are excellent sources of information on using non-antibody methods for identifying a wide range of organelles.

7.2.3 Mitochondria

If your staining pattern reveals large round or elongated structures (e.g. *Figure 4D*), then it may correspond to mitochondria. Mitochondrial morphology varies considerably, even from cell to cell, and methanol fixation often causes mitochondria to lose their elongated structure and become round. There are commercial antibodies to mitochondrial heat-shock proteins (see Appendix 2). Mitochondria are also easy to see using either phase contrast or DIC microscopy. Molecular Probes also sell a variety of vital dyes for mitochondria, some of which will remain after fixation. A cautionary note, however, is that mitochondrial labelling is the most common artefactual organelle-like staining. This can be due to autofluorescence following glutaraldehyde (and

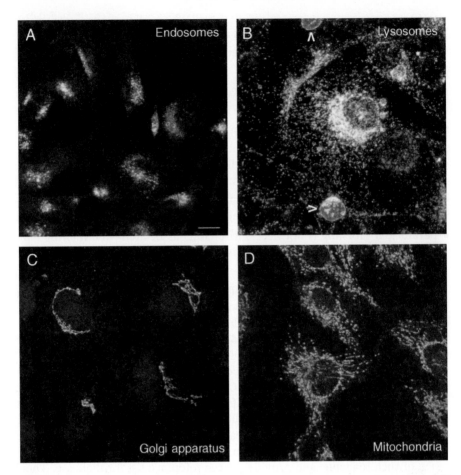

Figure 4. Immunofluorescent localization of a variety of organelle markers. (A and B) NRK cells labelled with antibodies to the mannose-6-phosphate receptor (A: an endosomal marker) and with antibodies to Lgp110 (B: a lysosomal marker). Labelled cells were provided by Dr Naomi Bishop, and the antibodies were from Dr Paul Luzio (University of Cambridge). Arrowheads in B mark artefactual blebbing present in some cells. (C) Vero (green monkey kidney) cells labelled with a polyclonal antibody to the Golgi apparatus protein GM130, provided by Dr Graham Warren, ICRF, London. (D) Mitochondria in *Xenopus* XL2 cells revealed using an antibody to HSP60 (Sigma, H-3524). Cells in panels A–C were fixed with formaldehyde and permeabilized with Triton X-100 (*Protocol 3*). Methanol fixation (*Protocol 2*) was used in panel D. Images were obtained using a Leica TCS NT confocal scanning microscope equipped with a krypton/argon laser, and are 2D projections of a z-series stack of eight images. Scale bar represents 10 μm for all panels.

sometimes methanol) fixation. In addition, since a number of fluorophore derivatives are used to label mitochondria (even in fixed cells), then it is possible that any free fluorochrome present in the secondary antibody may result in mitochondrial labelling.

Acknowledgements

I am grateful to the following people for providing images for figures: Emma Clarke (*Figure 3B* and *C*) and Jon Lane (*Figure 2C* and *D*). Thanks are also due to Naomi Bishop, Jon Lane and Shahida Rafiq for providing some of the specimens used in the figures. I would like to thank Pete Brown, Jon Lane and Philip Woodman for their comments on the text. Work in V.A.'s laboratory is funded by the Lister Institute of Preventive Medicine, the Wellcome Trust, the MRC and the BBSRC. V.A. is a Lister Fellow.

References

1. Harlow, E. and Lane, D. (1988) *Antibodies: A Laboratory Manual.* Cold Spring Harbor Laboratory Press, Cold Spring Harbor, NY.
2. Miller, D. M. and Shakes, D. C. (1995) In *Methods in Cell Biology* Vol. 48: *Caenorhabditis elegans: Modern Biological Analysis of an Organism* (eds H. F. Epstein and D. C. Shakes), p. 365. Academic Press, San Diego, CA.
3. Wick, S. M. (1993) In *Methods in Cell Biology*, Vol. 37: *Antibodies in Cell Biology* (ed. D. J. Asai), p. 171. Academic Press, San Diego, CA.
4. Bacallao, R., Kiai, K. and Jesaitis, L. (1995) In *Handbook of Biological Confocal Microscopy* (ed. J. B. Pawley), p. 311. Plenum Press, New York.
5. Wright, B. D. and Scholey, J. (1993) In *Methods in Cell Biology*, Vol. 37: *Antibodies in Cell Biology* (ed. D. J. Asai), p. 223. Academic Press, San Diego, CA.
6. Summers, R. G., Stricker, S. A. and Cameron, R. A. (1993) In *Methods in Cell Biology*, Vol. 38: *Cell Biological Applications of Confocal Microscopy* (ed. B. Matsumoto), p. 265. Academic Press, San Diego, CA.
7. Gard, D. L. and Kropf, D. L. (1993) In *Methods in Cell Biology*, Vol. 37: *Antibodies in Cell Biology* (ed. D. J. Asai), p. 147. Academic Press, San Diego, CA.
8. Gard, D. L. (1993) In *Methods in Cell Biology*, Vol. 38: *Cell Biological Applications of Confocal Microscopy* (ed. B. Matsumoto), p. 241. Academic Press, San Diego, CA.
9. Steuer, E. R., Wordeman, L., Schroer, T. A. and Sheetz, M. P. (1990) *Nature* **345**, 266.
10. Ellis, J. A., Jackman, M. R., Perez, J. H., Mullock, B. M. and Luzio, J. P. (1992) In *Protein Targeting: A Practical Approach* (eds A. I. Magee and T. Wileman), p. 25. IRL Press, Oxford.
11. Melan, M. A. and Sluder, G. (1992) *J. Cell Sci.* **101**, 731.
12. Hannah, M. J., Weiss, U. and Huttner, W. B. (1998) *Methods: A Companion to Methods in Enzymology* **16**, 170.

13. Griffiths, G. (1993) *Fine Structure Immunocytochemistry*. Springer-Verlag, Berlin.
14. Hayat, M. A. (1981) *Fixation for Electron Microscopy*. Academic Press, New York.
15. Harlow, E. and Lane, D. (1999) *Using Antibodies. A Laboratory Manual*. Cold Spring Harbor Laboratory Press, Cold Spring Harbor, NY.
16. McClean, I. W. and Nakane, P. K. (1983) *J. Histochem. Cytochem.* **22**, 1077.
17. Luther, P. W. and Bloch, R. J. (1989) *J. Histochem. Cytochem.* **37**, 75.
18. Luftig, R. B., McMillan, P. N., Weatherbee, J. A. and Weihing, R. R. (1977) *J. Histochem. Cytochem.* **25**, 175.
19. Tuma, M. C., Zill, A., Le Bot, N., Vernos, I. and Gelfand, V. (1998) *J. Cell Biol.* **143**, 1547.
20. Safiejko-Mroczka, B. and Bell, B. P. (1996) *J. Histochem. Cytochem.* **44**, 641.
21. Mitchison, T. J. and Kirschner, M. W. (1985) *J. Cell Biol.* **101**, 766.
22. Lindroth, M., Bell, P. B., Fredriksson, B. A. and Liu, X. D. (1992) *Microscop. Res. Tech.* **22**, 130.
23. Peränen, J., Rikkonen, M. and Kääriäinen, L. (1993) *J. Histochem. Cytochem.* **41**, 447.
24. Allan, V. J. and Kreis, T. E. (1986) *J. Cell Biol.* **103**, 2229.
25. Donaldson, J. G., Lippincott-Schwartz, J., Bloom, G. S., Kreis, T. E. and Klausner, R. D. (1990) *J. Cell Biol.* **111**, 2295.
26. Stearns, M. and Ochs, R. (1982) *J. Cell Biol.* **94**, 727.
27. Kreis, T. E. (1987) *EMBO J.* **6**, 2597.
28. Osborn, M. and Weber, K. (1982) *Methods Cell Biol.* **24**, 97.
29. Bonifacino, J. S., Klausner, R. D. and Sandoval, I. V. (1985) *Proc. Natl Acad. Sci. USA* **82**, 1146.
30. Stenmark, H., Vitale, G., Ullrich, O. and Zerial, M. (1995) *Cell* **83**, 423.
31. Heimer, G. V. and Taylor, C. E. D. (1974) *J. Clin. Pathol.* **27**, 254.
32. Johnson, G. D. and Nogueria Araujo, G. M. (1981) *J. Immunol. Methods* **43**, 349.
33. Johnson, G. D., Davidson, R. S., McNamee, K. C., Russell, G., Goodwin, D. and Holborow, E. J. (1982) *J. Immunol. Methods* **55**, 213.
34. Giloh, H. and Sedat, J. W. (1982) *Science* **217**, 1252.
35. Terasaki, M., Chen, L. B. and Fujiwara, K. (1986) *J. Cell Biol.* **103**, 1557.
36. Griffiths, G., Parton, R. G., Lucocq, J., van Deurs, B., Brown, D., Slot, J. W. and Geuze, H. J. (1993) *Trends Cell Biol.* **3**, 214.

2

Immunofluorescent labelling of sections

ALISON J. NORTH and J. VICTOR SMALL

1. Introduction

Over the past two decades, immunofluorescent labelling of cells and tissues has become one of the most powerful tools in cell biology. It is commonly performed in two ways: labelling of whole cells and labelling of sections. The problem with the first approach lies in rendering the intracellular targets accessible to the detecting antibodies. This necessitates some form of membrane permeabilization, generally using detergents or alcohols, or a direct physical approach such as micro-injection. Unfortunately such manipulations can lead to artefacts including ultrastructural changes and the loss or redistribution of the target antigen. Moreover, not all parts of the structure may be equally accessible to the antibody, even after extensive permeabilization. We have found this to be particularly pertinent in densely packed structures such as muscle (1). For this reason, the use of sectioned tissue, in which all cellular sub-structures are rendered unequivocally accessible at the section surface, can often prove to be the optimal procedure for consistent and reliable immunolocalization studies.

Immunofluorescent labelling is conventionally performed on sections between 5 and 10 μm in thickness. However, in this chapter we have largely concentrated on a less common procedure, namely the use of semi-thin (0.1–1 μm) sections. The immunolabelling procedures described towards the end of the chapter are equally applicable to both thick and thin sections. In an early review of his cryo-ultramicrotomy technique (2), Tokuyasu commented, 'I consider that semi-thin frozen sections, which allow the attainment of the highest resolution of light microscopy, are a new and powerful contribution to immunochemical studies of tissues and cells and that their specific advantages deserve a wider attention among cell biologists.' In this chapter we aim to highlight the importance and general applicability of semi-thin sections and to demonstrate the superior quality of images which may be obtained by this approach.

2. Choice of sectioning procedure

2.1 Conventional (5–10 μm) sections

Sections for immunofluorescent labelling are commonly obtained from frozen tissue (known as frozen or cryostat sections) or from paraffin-embedded tissue (paraffin sections). The advantage of frozen sections lies in the ease and speed of tissue preparation and sectioning, and in the lack of requirement for tissue fixation, with the result that the majority of antibodies will recognize their target antigen. The advantage of paraffin sections lies in their superior structural preservation, permitting correlation between labelling and morphology, and in the long-term stability of the specimens. However, paraffin embedding involves a number of preparative steps which can potentially affect antigenicity. Although antigenicity can sometimes be retrieved (see section 4), this often renders frozen sectioning the method of choice.

2.2 Semi-thin (0.1–1 μm) sections

2.2.1 The pros and cons

Semi-thin sections of tissue for immunofluorescent labelling are prepared in the same way as ultra-thin sections for electron microscopy. The major disadvantage of semi-thin sectioning is therefore the requirement for a specialized ultramicrotome. A further problem is the near invisibility of semi-thin sections under phase contrast, such that immunostaining cannot easily be correlated with cell or tissue architecture. This problem can be circumvented to some extent by nuclear counter-staining using Hoechst or propidium iodide (in the secondary antibody mixture or mounting medium; see Chapter 1, *Protocol 10*; *Figure 2c–g*; ref. 3). However, these disadvantages are outweighed by the many advantages of semi-thin sections, which include:

(a) Superior ultrastructural preservation: since semi-thin sections are produced under the same conditions as for immunoelectron microscopy. Ice crystal damage, one of the major causes of structural distortion in cryostat sections, is eliminated by the embedding and freezing protocol and the use of small tissue blocks. Furthermore, distortion through sectioning is minimized by the correct choices of embedding mixtures and cutting temperatures.

(b) Increased resolution: thus proteins can be assigned not only to a given cell type but also to specific compartments of the cell. The section thickness, typically 200–400 nm, is usually less than the diameter both of a single cell and of most of its membrane-bound compartments. This results in optimal accessibility of antigens within all internal structures (4) and a lack of superimposition of individual structures (*Figure 1*). Thus staining which may appear continuous on normal 5 μm sections can be resolved into discrete domains on semi-thin sections (*Figure 2a*; see also refs 5–10).

The resolution of nearby structures is increased by both the superior ultrastructure (see point (a) above) and the restriction of antibodies to the section surface (see section 5.3).

(c) The localization of several antigens within the same cell on serial sections; this is useful, for example, as an alternative to multiple immunolabelling when primary antibodies have been raised in the same species (11).

(d) The possibility of performing correlative immunofluorescence and immunoelectron microscopy (8,12,13).

(e) The ability to section very small pieces of tissue: this is clearly an advantage when only small pieces of tissue (such as punch biopsies) can be obtained, but can also be invaluable for studying embryonic organs. Since sectioned blocks are typically a 1 mm cube or less in size, minute structures can be embedded and sectioned whole (*Figure 2c–g*). The only alternative would be to use whole-mount staining and confocal microscopy, the disadvantages of which include the need for extensive permeabilization, long incubation times and an entire specimen per antibody.

(f) A requirement for only minimal amounts of antibody for immunolabelling, typically 5–10 µl per coverslip; this is an important point when antibody quantity is severely limited.

2.2.2 Methods of tissue preparation for semi-thin sectioning

Tissue can be prepared for thin sectioning and post-embedding immunolabelling by a number of methods (see ref. 14, for review). The most widely used approach has undoubtedly been the Tokuyasu cryo-ultramicrotomy procedure (15), in which the retention of cellular constituents in their hydrated state is highly favourable for immunolabelling. Immunofluorescent labelling has been performed on etched, Epon-embedded sections of freeze-dried tissue (6) but the harsh preparative procedures are often deleterious towards antigenicity and high background labelling can also be a problem (see ref. 2 and references therein). The methacrylate resins, such as the Lowicryl and London (LR White) resins, have also proved suitable embedding materials for immunofluorescence (16). These are more hydrophilic than conventional resins and the embedding procedure involves fewer deleterious steps, making them more generally applicable for immunocytochemistry. Reversible embedding procedures, in which the embedding matrix can be fully removed after sectioning to maximize antigen accessibility, are also potentially useful. Examples of these include infiltration of tissue with hard waxes, such as paraffin, polyethylene glycol or diethylene glycol distearate, or with polymethylmethacrylate (see ref. 17 and references therein). However, these embedding methods are generally complex and inapplicable to antigens that cannot tolerate exposure to organic solvents.

An alternative reversible embedding procedure utilizes the water-soluble resin, polyvinyl alcohol (PVA; 18). This method has principally been used for

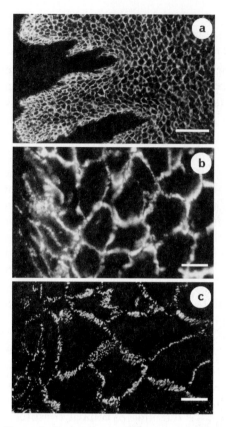

Figure 1. Immunolabelling of desmosomes (using anti-desmoplakin antibodies) on sections of bovine nasal epidermis. Panels (a) and (b) depict 7 μm cryostat sections, while (c) shows a 250 nm semi-thin cryosection. While the appearance of the cryostat section appears acceptable at low magnification (a), a higher magnification image (b) reveals considerable out-of-focus blur. In comparison, the punctate labelling of individual desmosomes is clearly resolved on the thinner section (c; photographed at the same magnification as (b)). Scale bars: (a) 100 μm; (b) and (c), 10 μm.

electron microscopy (1,10,18–20) but is equally applicable to fluorescence microscopy (16,18; A. J. North, unpublished data). PVA has many advantages over other embedding media, including its water-solubility, the ease of specimen preparation and its superior sectioning properties. The embedding procedure is very mild, with no requirement for dehydration through solvents, and the PVA can be removed from the specimen after sectioning simply by incubation on aqueous buffers. Furthermore, we have found PVA embedding to be applicable to a wide range of cell types, including cultured mammalian cells (*Figure 2e–g*), yeast cells and trypanosomes (A. J. North, unpublished data). Its only major disadvantage is its incompatibility with fatty tissues.

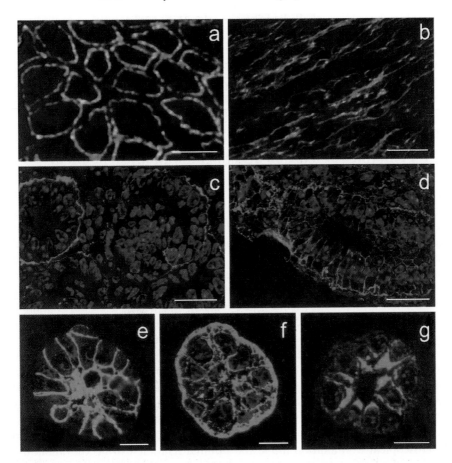

Figure 2. (a) Transverse semi-thin cryosection of smooth muscle labelled for vinculin (green) and dystrophin (red). Note the resolution of neighbouring membrane domains. (b) Artificially expanded semi-thin cryosection of smooth muscle labelled for desmin (red) and α-actinin (green). Expansion of the sections permitted the distribution of α-actinin-containing dense bodies along the desmin-containing cytoskeletal domains to be seen more clearly. (c and d) Semi-thin PVA sections of embryonic mouse kidney, fixed in methanol prior to embedding. The kidney was <1 mm³ in size and would therefore be unsuitable for cyrostat or paraffin sectioning. Note the contrasting distributions of laminin (c: green), surrounding epithelial tubular structures, and β1-integrin (d: green), along the basolateral surfaces of epithelial cells. The DNA counter-stain, Hoechst (blue), reveals the underlying morphology of the tissue. (e–g) Semi-thin PVA sections of a 'cyst' of MDCK cells, grown in methyl cellulose and fixed with methanol prior to embedding. Note the characteristic basolateral membrane-associated distribution of β1-integrin (e), the apical concentration of keratin (f), and the basal concentration of vimentin (g) (nuclei shown in blue). Scale bars: (a) and (b), 10 μm; (c) and (d), 20 μm; (e–g), 5 μm.

In this chapter we shall describe the preparation of semi-thin sections by both the Tokuyasu method and the PVA-embedding method. We find that both of these methods give excellent immunofluorescent labelling for the majority of antigens tested. The choice between them will depend on the type of tissue or cells to be sectioned, the type of equipment that is available, and the resistance of the antibody to the different preparative steps.

3. Tissue fixation

3.1 Principles of fixation and the immunocytochemical compromise

Tissue fixation is usually necessary to preserve the ultrastructure of the cell and to prevent redistribution or loss of cellular components. Immunolabelling can be performed on unfixed conventional cryostat sections; indeed, this is generally the method of choice for antibodies that cannot withstand chemical fixation. However, many components leach out of unfixed sections, so that labelling of certain antigens will require fixed tissue. Moreover, unfixed sections will deteriorate rapidly after labelling and thus cannot be stored long term. Therefore, cryostat sections are typically fixed using acetone, methanol, or a combination of the two, at a temperature of –20°C to minimize denaturation of precipitated proteins (see refs 21 and 22 for reviews).

For higher resolution immunocytochemical studies or even for paraffin sections, the compromise between ultrastructural preservation and retention of antigenicity can be a major problem. A further important consideration is the accessibility of antigen to antibodies. Apparent loss of antigenicity may often be caused by extensive cross-linking of the specimen, preventing access- ibility, rather than by chemical alteration of the antigen. Therefore for light microscopical examination alone the optimal fixation will be the minimum necessary to preserve the integrity of the target structure and prevent re- distribution of the antigen (2). Hence milder fixatives can be employed for immunofluorescence than would be recommended for immunoelectron microscopy. We typically use a brief fixation (0.5–1 h at room temperature or 4°C) in 2% (w/v) buffered formaldehyde, although even shorter times (such as 5 min) may be better for particularly sensitive antibodies.

When correlative fluorescence and immunoelectron microscopy are to be performed, the ultrastructure can be significantly improved by the addition of low concentrations of glutaraldehyde (0.001–0.1% v/v). Moreover, some antigens absolutely require the use of glutaraldehyde fixation. We recom- mend testing the fixation sensitivity of each antibody on cryostat sections before proceeding to semi-thin sectioning. To accomplish this, cryostat sections on slides are fixed for 5 min in a range of fixatives such as acetone, methanol, ethanol, formaldehyde, or formaldehyde plus glutaraldehyde and subsequently stained by immunofluorescence. The strongest fixative that still permits intense labelling is then selected.

3.2 Preparation of buffered fixative solutions

Fixative solutions are made according to the general method of Karnovsky (23). The basis for these fixatives is an aqueous stock solution of formaldehyde, freshly made from paraformaldehyde powder to avoid contaminants such as methanol and formic acid. An optimal buffer, adjusted to the correct pH and osmolarity, should be chosen for each tissue to be studied (22) although 0.1 M phosphate buffer, pH 7.2, has been specified in most of the protocols below for the sake of simplicity. For immunofluorescence, however, the buffer is less critical than for immunoelectron microscopy.

Protocol 1. Preparation of aldehyde fixatives

Equipment and reagents
- Paraformaldehyde powder (Agar Scientific Ltd)
- 1 M NaOH
- Appropriate buffer, at double the usual concentration[a]
- Glass flask, funnel and Petri dish
- 25% or 50% (v/v) glutaraldehyde solution (EM grade, vacuum distilled; Agar Scientific Ltd)

Method

1. Add 8 g of paraformaldehyde powder to approximately 100 ml double-distilled H_2O in a glass flask. Cover the flask with a glass funnel and an inverted glass Petri dish, forming a reflux system.

2. Heat the solution, with stirring, at 60–70°C for at least 30 min, until the paraformaldehyde has been largely reduced.[b]

3. Clear the remaining cloudiness by adding a few drops of NaOH.

4. Once cool, filter through filter paper and make up the stock solution to 200 ml by adding double-distilled H_2O. Combine equal volumes of fixative with 2× buffer, to produce a 2% (w/v) formaldehyde solution in buffer.[c]

5. If required, add a suitable volume of glutaraldehyde to the buffered formaldehyde immediately before use.

6. Readjust the pH to around 7.2 when necessary.

[a] See section 3.2 above for choice of buffer.
[b] Perform this step in a fume cupboard.
[c] For stringent ultrastructural studies fixatives should be made up fresh. When processing tissue for immunofluorescence alone, fixatives can often be frozen either in water or in buffer as conveniently sized aliquots, and stored at –20°C. Upon subsequent thawing, formaldehyde concentrations of ≥4% (w/v) may require heating to 50°C to re-dissolve.

After aldehyde fixation, treatment of the tissue with sodium borohydride to inactivate any unbound aldehyde groups is advisable. This is particularly important when glutaraldehyde (a dialdehyde) is used, and also when embedding in PVA, as this would otherwise react with free aldehyde groups, preventing its subsequent extraction from the section.

3.3 Fixation and processing for aldehyde-sensitive antigens

Due to their fragility, it is difficult to cut semi-thin cryosections of tissue without prior aldehyde fixation. However, tissue fixed using methanol alone can be embedded in PVA and processed for semi-thin sectioning (*Figure 2c–g* and ref. 20). For antigens which cannot withstand any chemical fixation, semi-thin sections may be prepared from tissue which has been rapidly frozen, freeze-dried and embedded in Epon (6) or in Lowicryl (24).

4. Preparation of conventional (5–10 μm) sections

The preparation of cryostat and paraffin sections has been covered by a number of excellent and detailed reviews (see ref. 25 and references therein) and so will not be described here. While such sections are conventionally 5–10 μm thick, thinner (1–2 μm) paraffin sections can be obtained from blocks cooled to –20°C, while addition of 20% (w/v) sucrose to the OCT (optimal cutting temperature) embedding compound permits the cutting of 3 μm cryostat sections which give considerably improved resolution over 10 μm sections (26).

Various methods using proteolytic pre-digestions or microwave heating have been used to unmask epitopes in paraffin sections that had been rendered inaccessible by formalin-induced protein cross-linking (27,28). Results obtained with different pre-treatments vary greatly with the antibodies used (29).

5. Preparation of semi-thin (0.1–1 μm) sections

5.1 Processing tissue

Tissue should be dissected into appropriately sized pieces prior to fixation, using a dissecting microscope at all stages to achieve the desired orientation in the final mounted block. Specimens that need to be maintained at a particular length during fixation (such as muscle) should be pinned out for the first stage of fixation and then dissected into smaller pieces to complete the fixation process. A final size of $1 \times 1 \times 1$–3 mm is suitable for most specimens. After fixation the tissue is washed several times in buffer. Formaldehyde fixation is reversible, so if the tissue must be left in buffer for more than a few hours at this stage it is advisable to include a low concentration (0.3–1% w/v) of formaldehyde (15). However, washing the specimen in buffer alone overnight at 4°C has been reported to improve accessibility for immunolabelling, presumably because of this reversal of cross-linking (30).

To obtain a coherent section from loosely associated or fragile tissue, embedding the tissue in 10% (w/v) gelatin may prove helpful (31), though we have found that this sometimes reduces the labelling intensity. Apparently gelatins from different sources differ widely in their sectioning properties: Merck gelatin (no. 4078, with grain size 800 μm) has been reported to be easy to section and to result in vivid immunolabelling (32). A method of embedding embryonic tissues in polyacrylamide gel has also been described (33).

Protocol 2. Embedding tissue in gelatin

Equipment and reagents
- 10% (w/v) gelatin in 0.1 M phosphate buffer pH 7.2

Method

1. Immerse fixed and rinsed tissue pieces in 10% (w/v) gelatin at 37 °C for a few minutes.

2. Cool to 4°C to allow the gelatin to solidify, forming a thin (~1.5 mm) slab.

3. Cut small pieces of embedded tissue and re-fix for 30 min.

4. Process for cryo-ultramicrotomy (*Protocol 5*).

5.2 Processing single cells or cellular aggregates

Single cells and aggregates or monolayers of cells can also be embedded and sectioned by ultramicrotomy (*Figure 2e, f* and *g*). To embed single cells for semi-thin cryosectioning, cells are fixed in suspension, washed and then pelleted into fluid gelatin or agarose which is subsequently allowed to solidify prior to freezing and sectioning. Ideally a large, dense pellet should be used, so that cells can easily be found in all sections. If the pellet is small and difficult to find, Sephadex beads that have been stained using Evans Blue (a 2% (w/v) aqueous solution) can be included as markers that are visible in both frozen blocks and solutions.

Protocol 3. Agarose embedding of cells for ultramicrotomy[a]

Equipment and reagents
- 2% (w/v) solution of low melting point agarose in PBS

Method

1. Liquify the agarose solution by heating to 40 °C.

2. Pellet the fixed and washed cells in a microcentrifuge tube.

Protocol 3. *Continued*

3. Resuspend the cells in the warm agarose.

4. Pellet the cells, carefully ensuring that the agarose does not set before the pellet forms.[b]

5. Place the tube on ice for 5 min, to allow the pellet to solidify.

6. Remove the cells in agarose and cut into small pieces (c. 1 mm cubes).

7. Process the blocks for cryo-ultramicrotomy (see below)

[a] Based on methods described in ref. 34 and modified by P. M. Lackie (personal communication).
[b] This can be achieved by placing the small tube of cells plus agarose into a larger tube of warm water.

Cell monolayers can be scraped off the tissue culture dish and gently pelleted before processing as for single cells. Alternatively, the introduction of structural distortions during centrifugation can be avoided by growing monolayers on polycarbonate filters (Costar) and embedding the cells and filter together. The filters are fixed, washed, infiltrated with cryoprotectant (PVP/sucrose; see section 5.5 and *Protocol 5*, step 5), and cut into small strips which are mounted onto a cryopin and frozen for sectioning.

To embed cells in PVA, a pellet of fixed and washed cells is gently eased from a cut microcentrifuge tube into the bottom of a dish and surrounded with PVA, which will gradually infiltrate the cells before hardening. An area of PVA containing densely packed cells can be selected for sectioning using phase contrast microscopy.

5.3 Section thickness

One explanation for the relatively infrequent use of semi-thin sections may be the common assumption that the intensity of immunostaining is proportional to section thickness. However the evidence in favour of this assumption is highly controversial. Using a quantitative method for analysis of immuno-fluorescence staining, Mosedale and colleagues (35) reported that increasing the section thickness merely increased autofluorescence, with little effect on specific staining. Moreover, only marginal differences in labelling intensity were observed between a re-extractable medium (PVA) and non-extractable media (Lowicryl and LR White; 16). Studies employing 'transverse' ultra-microtomy of re-embedded immunolabelled ultra-thin sections have demonstrated the relative impermeability of cryosections to antibodies, except at regions of structural damage (36,37). Thus it appears that fixative-induced cross-linking may restrict the penetration of antibodies into tissue, with the result that there is no clear advantage for immunolabelling to using a thicker section.

We find a semi-thin section thickness of 0.15–0.25 μm to be optimal for most purposes. With thicker sections both resolution and section spreading are compromised, while the signal from nuclear stains (e.g. Hoechst) or other specific stains, such as phalloidin, will be weak in sections thinner than 0.15 μm. *Figure 2a* shows a 0.25 μm semi-thin cryosection of smooth muscle which has been double-labelled using antibodies against vinculin (green) and dystrophin (red). The result permitted assignment of these proteins to alternating membrane domains, each approximately 0.5 μm wide. However, resolution of these neighbouring domains could not be achieved in sections of ≥1 μm in thickness, each antibody label often appearing continuous rather than punctate.

Still greater resolution of neighbouring cellular regions has been achieved in artificially expanded sections (*Figure 2b*; see also refs 1 and 19). These are obtained by transferring sections (using a wire loop) onto buffer, prior to their retrieval. Although the integrity of the section is clearly compromised, this process can be useful for separating closely apposed structures.

5.4 Handling semi-thin sections

Semi-thin sections can be retrieved onto glass microscope slides, but given their minute size they can be difficult to locate unless their position is carefully noted. We find it more convenient to retrieve them onto 4 mm square glass coverslips, which are simply broken from a large coverslip that has been scored using a diamond pencil. To prevent sections from falling off during the immunolabelling procedure the coverslips are cleaned and coated with poly-L-lysine (see *Protocol 4*).

Protocol 4. Poly-L-lysine coating of coverslips

Equipment and reagents

- 4 mm glass coverslips
- Whatman no. 1 filter paper
- Acid/alcohol cleaning solution (60% (v/v) concentrated HCl and 40% (v/v) ethanol)
- Bacitracin (Sigma), 80 μg/ml aqueous solution
- Poly-L-lysine (Sigma), 1 mg/ml aqueous solution

Method

1. Clean the coverslips by gentle agitation for a few minutes in acid/alcohol cleaning solution.

2. Wash the coverslips with several changes of double distilled water.

3. Dry the clean coverslips and store on filter paper in Petri dishes.

4. Using anti-capillary forceps, dip each coverslip briefly into bacitracin solution, then drain off the excess against a torn edge of filter paper.

5. Dip the coverslip into poly-L-lysine, place it on filter paper, then turn it over, thus spreading the poly-L-lysine evenly across both surfaces.

6. Store coated coverslips on filter paper in Petri dishes.

5.5 Sectioning by the Tokuyasu method

The cryo-ultramicrotomy technique pioneered by Tokuyasu and others has been described in many excellent reviews (2,14,15,33,38) to which readers may wish to refer for a detailed consideration of the technique. A generalized protocol derived from these observations is presented below.

Largely accepted to be the most sensitive preparative method for immuno-electron microscopy, the major disadvantage of the Tokuyasu technique is the need for a cryo-ultramicrotome. However, when one is available, the technique is consistent, applicable to most tissues and antibodies (with the general exception of those which cannot withstand aldehyde fixation), and rapid: tissue can be fixed, infiltrated, sectioned and labelled in one or two days. Although cryosectioning has the reputation of requiring great skill and technical expertise, the design of modern cryo-ultramicrotomes has considerably improved the ease of cutting and the reproducibility of the sections. Certainly semi-thin sectioning presents no difficulties.

5.5.1 Cryoprotection

After fixation and washing the tissue is infiltrated with a concentrated sucrose solution or a mixture of sucrose and polyvinylpyrrolidone (PVP) in order to achieve cryoprotection during the subsequent freezing step and to confer plasticity upon the block. The addition of PVP confers greater plasticity than sucrose alone (39) and is particularly recommended for semi-thin sectioning.

5.5.2 Mounting and freezing

Infiltrated specimens are further trimmed (we recommend the use of fine commercial razor blades at this stage, rather than the single-edged industrial ones) before mounting onto cryopins (Leica). This minimizes the amount of trimming which will need to be performed in the microtome. The tissue piece should be orientated carefully on the cryopin to give the desired plane of section, and supported on either side by sucrose/PVP. After plunge-freezing into liquid nitrogen the specimens can be transferred, still under nitrogen, directly into the chamber of the cryo-ultramicrotome or can be stored for several months under liquid nitrogen without any noticeable deterioration.

5.5.3 Sectioning

Cryosectioning is performed as described previously (2,14,33,40). Once the temperature of the block has equilibrated to that of the microtome, the block face is trimmed to a small rectangle using the corner of a glass knife or a Leica trimming device. A suitable temperature for trimming PVP-infiltrated specimens is –65 to –75°C: for semi-thin sectioning the temperature is lowered to between –75 and –90°C. The exact temperatures, section thickness setting and cutting speed must be optimized for each tissue. Sectioning is best performed on dry glass knives. We find it convenient to move sections (using an eyelash) away from the knife edge into groups of four or five, ready for retrieval.

5.5.4 Section retrieval

Sections are conventionally retrieved using a wire loop containing a droplet of 2.3 M sucrose or 2 M sucrose plus 0.75% (w/v) gelatin (15). More recently, droplets of sucrose/methylcellulose or methylcellulose/uranyl acetate mixtures have been reported to improve section integrity (41). The droplet is lowered to the sections before it freezes, then removed from the cryochamber and warmed to room temperature. The reduced surface tension allows the sections to flatten as they thaw out. Sections are then transferred to coated coverslips and stored for up to one day on gelatin plates prior to immunolabelling.

Protocol 5. Preparation of ultra-thin cryosections by the Tokuyasu method

Equipment and reagents

- Cryo-ultramicrotome
- Cryopins (Leica)
- 0.1 M phosphate buffer pH 7.2
- PVP/sucrose solution: make a paste of 20 g of polyvinylpyrrolidone (PVP-10; mol. wt 10000; Sigma), 4 ml of 1.1 M Na$_2$CO$_3$ and 80 ml of 2 M sucrose in buffer. Cover the container and leave it at room temperature overnight; during this time, minute bubbles will escape from the paste, leaving behind a clear solution

- Aldehyde fixative (see *Protocol 1*)
- Sodium borohydride dissolved in phosphate buffer, immediately before use, to a concentration of 0.5 mg/ml
- Sucrose/gelatin solution: 2 M sucrose plus 0.75% (w/v) gelatin in PBS)
- Coated coverslips
- Gelatin plate: 35 mm dish half-filled with a warm solution of 2% (w/v) gelatin in PBS and left to solidify at 4°C

Method

11. Cut freshly obtained tissue into small pieces and fix them by the appropriate method (see section 3).

12. Wash with buffer four times for 10 min each.

13. Incubate with three successive changes of 0.5 mg/ml sodium borohydride solution, each for 5–10 min.

14. Wash three times in buffer, for 5 min in each change.

15. Infiltrate with PVP/sucrose solution, for 2–3 h at room temperature or overnight at 4°C, preferably with gentle agitation.

16. Mount the specimens on cryopins and freeze them immediately by rapidly plunging them into liquid nitrogen with continual agitation.

17. Cut semi-thin sections using a cryo-ultramicrotome, with a thickness setting of around 250 nm and temperatures around –85°C.

18. Retrieve sections on a droplet of sucrose/gelatin, allow them to thaw, and transfer them to coated coverslips.

Protocol 5. *Continued*

19. Place the coverslips, section side down, on a gelatin plate to be stored at 4 °C prior to immunolabelling.

10. Immediately before immunolabelling, fluidify the gelatin by incubation in a 37 °C oven, leaving the coverslips on the liquid gelatin for a further 10 min before beginning the immunolabelling procedure.[a]

[a] This procedure allows the coverslips to be lifted off the gelatin without tearing the sections and acts as a first blocking step against non-specific antibody binding. Incubation on 2% (w/v) gelatin is also reported to reduce the non-specific background staining of fluorescent molecules on glass (30).

5.6 Sectioning of PVA-embedded material

The advantages of the PVA technique were described above (section 2.2.2). In addition, PVA sections are less fragile than cryosections so that a milder fixation protocol, such as ice-cold methanol for 10–30 min, may suffice. Although only the most mechanically resilient structures will be preserved sufficiently well for ultrastructural studies (20), the morphological preservation is generally adequate for light microscopical analysis. Both the embryonic kidneys and the MDCK cysts shown in *Figure 2* were fixed using methanol alone. The blocks can be stored for months or even years, permitting the same sample to be resectioned and labelled with additional antibodies later. After prolonged storage the blocks may become too dehydrated for immunolabelling with some antibodies, but this can be reversed by placing them in a humidified atmosphere for a few hours.

5.6.1 Embedding in PVA

Embedding in PVA is extremely simple, the fixed and washed specimens (after sodium borohydride treatment) merely being immersed in a dish of 20% (w/v) aqueous PVA. Hardening of the PVA can be achieved overnight at 60 °C, but this abolishes the antigenicity of some proteins. A method that is compatible with more antigens is simply to leave the dish open at room temperature for about 3 weeks. Once hard, blocks of embedded tissue or cells are then cut out of the dish using a razor blade. Certain antibodies will not label totally dried specimens. In this case the blocks can be partially dried to a jelly-like consistency and sectioned at low temperatures (10), but this sectioning procedure is technically more demanding.

5.6.2 Sectioning

PVA sections are trimmed and cut in the same way as for epoxy sections but using glycerol as the flotation medium (18). The crumpled appearance of newly cut sections can be disconcerting, but as the sections move further down the boat they gradually spread until they are flat with classical interference colours.

Protocol 6. Preparation of sections of PVA-embedded material[a]

Equipment and reagents

- Ultramicrotome
- Aldehyde fixative (see *Protocol 1*) or –20°C methanol
- 87% (v/v) glycerol (aqueous solution)
- 50% (v/v) glycerol (aqueous solution)
- 0.1 M phosphate buffer, pH 7.2
- PVA solution: 20% (w/v) aqueous solution of polyvinyl alcohol 203, 10000 mol wt (Air Products), cleared of undissolved material by centrifugation at 16000*g* for 30 min

Method

Process the tissue as for steps 1–4 of *Protocol 5*[b] before continuing with the following procedures.

11. Place the tissue in a 35 mm plastic dish containing PVA solution (one-third to half full), and agitate gently for 2–3 h when appropriate.[c]

12. Leave the dish uncovered at room temperature for around 3 weeks or until the PVA has hardened completely. Top up with additional PVA as it dries to ensure that the embedding layer is at least as thick as the tissue block.

13. Cut out small rectangular regions of tissue embedded in PVA using a fresh razor blade.

14. Mount the specimen in an ultramicrotome chuck, using a small piece of cardboard on either side to support the PVA block.

15. Trim the block face using a glass knife to obtain a small square or rectangular face (<0.5 mm across).

16. Cut sections using a cutting speed of around 1 mm/sec and a thickness setting of 1.2–1.6 μm. Float the sections onto a boat of 87% (v/v) glycerol, maintaining the level of fluid just below the knife edge.[d]

17. After sectioning, allow the sections to spread by introducing a few drops of 50% (v/v) glycerol into the boat.

18. Fill the boat by addition of 87% (v/v) glycerol.

19. Retrieve the sections onto coated coverslips by contact from above.

10. Store sections for up to a few days at 4°C inverted on a dish of 87% (v/v) glycerol.

11. Prior to immunolabelling, extract supporting PVA from the section by incubation on buffer for a few hours at room temperature or overnight at 4°C.

[a] Modified from the method described in ref. 18 and modified in ref. 20.
[b] PVA embedding is not compatible with a high lipid content in the tissue, so fatty parts of the tissue should preferably be removed from the region of interest (such as dermis from epidermis) at this stage.
[c] Agitation is recommended for solid pieces of tissue but not for loose pellets of cells.
[d] PVA, being a water-soluble resin, is strongly hydrophilic. A low fluid level and the use of a diamond knife are therefore recommended in order to keep the block face dry. If the block face becomes wet, dry the block immediately in a 37°C oven and then re-trim the surface using a dry glass knife, before resuming sectioning.

6. Immunofluorescent labelling

6.1 Labelling procedure

Conventional sections on slides are normally labelled in a humidified slide chamber, with washes being performed in racks in glass dishes or in Coplin jars. Immunolabelling of semi-thin sections is performed by transferring coverslips from one droplet of solution to the next, across a clean sheet of Parafilm. This convenient method miminizes the amount of each antibody needed as the weight of the inverted coverslip spreads the antibody solution evenly beneath the section or cells and prevents it from drying out too quickly. Much larger droplets are used for washes.

Four-millimetre coverslips can be rapidly transferred between droplets using a 3–4 mm diameter wire loop mounted on a wooden stick. When transferring onto antibody solution, however, it is better to use a pair of anti-capillary forceps and to drain excess buffer from one side of the coverslip (without allowing the sections to dry out), in order to prevent dilution of the antibody. A plastic dust cover is placed over the drops during incubations, taking care to include some larger droplets of PBS alongside in order to provide an humidified atmosphere. After labelling, each coverslip is inverted onto a small drop of mountant on a glass slide and allowed to dry before examination or temporary storage in the dark at 4°C.

6.2 Primary and secondary antibodies

Immunolabelling is generally performed using an indirect method, in order to provide signal amplification and to permit the use of commercial fluorochrome-conjugated secondary antibodies. Monoclonal antibodies often give the most specific labelling, but polyclonal antibodies, particularly after affinity purification, can be more useful for localizing antigens in fixed tissue as it is unlikely that all epitopes will be altered or obscured. Many companies produce excellent secondary antibodies: we have found that the Jackson ImmunoResearch 'Multiple Labelling' antibodies give clean, bright staining on most specimens.

If the signal is weak, further amplification steps, such as the use of a bio-tinylated secondary antibody and a streptavidin-conjugated fluorochrome, or a fluorochrome-conjugated tyramine (42), may prove beneficial. An alternative is to use a low light level CCD camera to enhance faint signals (43). Such a system also permits the use of fluorochromes whose emission spectra lie in the far-red region, such as cyanine 5 (Cy5), thereby reducing the problem of bleed-through between signals.

6.3 Optimization of labelling conditions

Parameters such as antibody concentration, length of antibody incubation and incubation temperature should ideally be optimized for each primary antibody used. It is advisable to start with a simple labelling procedure (see *Protocol 7*) and to modify it as appropriate if the staining is weak or a high background of non-specific label is obtained. A reasonable starting point for optimization is a primary antibody concentration of 10–25 µg/ml incubated for 30–60 min at room temperature or 37 °C (44,45). However, longer incubation times in both primary and secondary antibodies have been reported to increase staining intensity and different primary antibodies bind optimally at different temperatures, between 4 and 37 °C (35). We have found that the intensity and specificity of labelling seen after a 1 h incubation at room temperature can sometimes be improved by an overnight incubation at 4 °C using a lower concentration of antibody.

Immunolabelling is sometimes improved by the addition of a low percentage of detergent (such as Triton X-100, Tween 20 or Nonidet P-40) to the primary antibody and/or washes, but this is again highly dependent on the individual antibody.

6.4 Controls

It is important to include both positive and negative controls to permit any conclusions to be drawn from labelling using either a new antibody or tissue prepared by a different method. Commonly performed negative controls include:

- no primary antibody (to detect non-specific binding of the secondary antibody);
- no primary or secondary antibody (to identify any tissue autofluorescence— see section 7);
- replacement of the primary antibody with pre-immune serum (for polyclonal antisera) or an irrelevant monoclonal antibody (directed against an antigen that is not expressed in the tissue);
- the use of primary antibody which has been cross-adsorbed with its antigen;
- labelling of a tissue negative for the antigen.

It is equally important to perform suitable positive controls, such as parallel labelling using an antibody with a known and specific labelling distribution on tissue prepared by the chosen method. This demonstrates the specificity of the antibody under test and also allows the general reactivity of the tissue sections and of the secondary antibody to be assessed.

6.5 Strategies for multiple immunolabelling

Immunolabelling of more than one antigen is simple if the primary antibodies were raised in different species. Two or more such antibodies can be applied simultaneously, as can the secondary antibodies conjugated to different fluorochromes. Care must be taken when choosing the fluorochromes (see

Appendix 1) that the microscope has appropriate combinations of filters to prevent bleed-through of one signal into the other channels. Since the advent of monoclonal antibodies, however, the need to label with two antibodies from the same species has become increasingly common. Conjugation of each primary antibody to a different fluorochrome is possible but direct labelling often results in a weak signal. A number of other options to circumvent the problem of cross-reactivity are:

(a) The first mouse monoclonal antibody is applied, then the sections are blocked with a non-binding mouse antibody together with unconjugated goat anti-mouse IgG F(ab) fragments, before application of the second primary and secondary antibodies (46).

(b) The mouse monoclonals and their respective conjugated anti-mouse antisera are first allowed to complex *in vitro* and the unbound secondary antibodies are blocked with normal mouse serum. Upon applying this solution to the sections the antigens can be recognized only by the immune complex (47).

(c) Microwave oven treatment of sections between sequential rounds of immunostaining can block antibody cross-reactivity. This method also permits the use of sensitive three-layer staining for each primary antibody (48).

(d) Biotinylation of one of the primary antibodies permits its recognition using an avidin- or streptavidin-conjugated fluorochrome. Then blocking with excess normal serum of the same species as the primary antibodies is needed between the two cycles of staining (49).

(e) The first staining pattern is eliminated after photography by chemically removing the antibodies, prior to reincubation with a second antibody (6).

Protocol 7. Immunolabelling of sections

Equipment and reagents

- 0.02 M glycine
- Blocking solution: 1% (w/v) bovine serum albumin (BSA) plus 5% (v/v) normal goat serum (NGS) in PBS
- Primary antibody, diluted as appropriate using 1% (w/v) BSA in PBS
- 0.1% (w/v) BSA in PBS
- Secondary antibody: generally used at a dilution of 1/100 or lower and pre-incubated with 1% (w/v) BSA plus 10% (v/v) serum (preferably from the same species as the tissue section) for 1 h on ice, to minimize non-specific binding

- Gelvatol: add 0.136 g K_2HPO_4 to 100 ml of water and adjust to pH 7.2 using NaOH solution. To 40 ml of this solution add 10g Gelvatol (PVA, 10000 mol. wt; Sigma) and stir overnight to dissolve. Add 20 ml of 99% (v/v) glycerol and continue to stir until homogeneous. Centrifuge for 15 min at 12000 r.p.m. (approx 20000 g) to clear, then adjust the pH to around 7. Store at 4°C
- Paraphenylene diamine (PPD; Sigma) dissolved at 10 mg/ml in water, then stored as 20 μl aliquots at –20°C

Method

1. Incubate the sections with three changes of 0.02 M glycine, for a total of 10 min.[a]

2. Block the sections for 10 min using blocking solution.[b,c]

3. Apply primary antibody to the sections for the appropriate time (such as 1 h) at optimal concentration (section 6.3).

4. Transfer the sections across five washes of PBS/0.1% (w/v) BSA for a total of 10 min.

5. Incubate the sections with secondary antibody for 30 min at room temperature.

6. Transfer the sections across five washes of PBS for a total of 20 min.

7. Immediately before mounting, gently mix 180 μl of Gelvatol with one aliquot (20 μl) of PPD, avoiding the introduction of air bubbles into the solution.[d]

8. Mount the sections in the Gelvatol/PPD mixture.

9. Dry the preparations for 2 h at room temperature or overnight at 4°C; store them in the dark at 4°C and examine them as soon as possible.

[a] Glycine inactivates free aldehyde groups within the section, so this step is only necessary for aldehyde-fixed specimens.
[b] Ideally the source of serum should be the same species as that of the secondary antibody.
[c] If non-specific labelling remains after blocking with NGS and BSA alone, other blocking agents can be added to the blocking solution such as 2% (w/v) gelatin and/or 2% (w/v) PVA.
[d] Do not store the PPD pre-mixed with Gelvatol, as this mixture will not freeze at −20°C and the anti-fade will deteriorate (see section 8).

7. Autofluorescence

Autofluorescence is caused by certain biomolecules, such as fibronectin, elastin and lipofuscin, and also by aldehyde fixation (50). It is thus more prominent in certain tissues than in others, and can be particularly problematic if the source of autofluorescence lies close to a weak immunocytochemical signal, although this difficulty can sometimes be overcome by using confocal microscopy to examine the sections (49). Since the range of the spectra of autofluorescent molecules is very broad, autofluorescence cannot simply be eliminated by switching from one fluorescent probe to another with different excitation and emission spectra. However, the problem can often be reduced by this means, as the use of fluorophores excited by longer wavelengths such as Texas Red or rhodamine (Appendix 1) typically provide less spectral overlap with endogenous fluorescent molecules. Since the intensity of autofluorescence is related to section thickness (35,51), it poses less of a problem with semi-thin sections.

A number of methods have been applied to reduce or eliminate autofluorescence. A low-fading contrasted immunofluorescence method, in which the sections were mounted in a medium containing PPD and propidium iodide, was reported to reduce autofluorescence even after fixation in formaldehyde plus glutaraldehyde (3). Autofluorescence can be eliminated by the use of digital image processing to subtract the autofluorescence image, captured at an excitation wavelength outside the excitation spectrum of the fluoroprobe (51). A more sophisticated method to eliminate autofluorescence is time-resolved fluorometry, but this requires a specialized and expensive fluorescence microscope (52).

8. Anti-fading mountants

Photobleaching of fluorescent staining can be a significant drawback of the technique if not addressed. The fluorescein dyes are particularly susceptible to rapid fading, but there is some discrepancy as to whether the rhodamine dyes are also subject to fading (53,54). Clearly, the degree of photobleaching will also depend on the imaging system. Thus the problem is more significant for confocal microscopy than for low light level microscopy.

A number of studies have compared the use of several mounting media containing different anti-fading agents. Most were found to induce a slower fluorescence decay but they also significantly quenched the initial fluorescence intensity. PVA-based mounting media have been found to give a higher intensity of staining and lower background fluorescence than buffered glycerol (54), although the pH of the PVA mountant needed to be re-adjusted regularly. Three commonly used anti-fading agents are PPD, *n*-propyl gallate (NPG) and 1,4-diazobicyclo(2,2,2)-octane (DABCO). Several groups have reported PPD to give the best retardation of fading (54,55), although NPG gave the best quality of preparations and therefore was excellent for bright staining (55). The only advantage of DABCO was a better stability on storage (54). Specimens mounted in PVA mountant containing PPD should be examined within a few days because a progressive discoloration from the edges of the coverslips is induced by PPD (54). (This is especially a problem when using 4 mm coverslips.) The authors therefore recommended remounting in PVA alone for prolonged storage of the tissue sections. Of the commercially available anti-fading preparations, Vectashield was commonly recommended (55,56).

9. Concluding remarks

The techniques described above provide results that are comparable and often superior to those obtained by confocal microscopy of thicker specimens, with the added advantage of correlative light and electron microscopy. It is clear, however, that the chosen route will depend on equipment availability. Cryo-ultramicrotomy remains a fairly specialized technique, although its technical difficulties have been reduced by the development of more user-friendly equipment. It is clearly important to seek new embedding media for room temperature sectioning, with improved handling and sectioning characteristics, greater stabilization of tissue ultrastructure and a more universal retention of antigenicity. Resins that are compatible with hydrated specimens are an obvious route forward. Another way may be to use mixed embedding media, such as the combination of agarose and acrylamide which has proved successful in embedding embryos for thick sectioning (57). The possibility of combining such mixed embedding media with subsequent freezing or PVA infiltration to permit ultramicrotomy remains to be explored.

References

1. North, A.J., Gimona, M., Cross, R.A. and Small, J.V. (1994). *J. Cell Sci.*, **107**, 437.
2. Tokuyasu, K.T. (1980). *Histochem. J.*, **12**, 381.
3. Jensen, H., Broholm, N. and Norrild, B. (1995). *J. Histochem. Cytochem.*, **43**, 507.
4. Sander, H.J., Slot, J.W., Bouma, B.N., Bolhuis, P.A., Pepper, D.S. and Sixma, J.J. (1983). *J. Clin. Invest.*, **72**, 1277.
5. Maher, P.A., Cox, G.F. and Singer, S.J. (1985). *J. Cell Biol.*, **101**, 1871.
6. Drenckhahn, D. and Franz, H. (1986). *J. Cell Biol.*, **102**, 1843.
7. Geerts, A., Geuze, H.J., Slot, J.-W., Voss, B., Schuppan, D., Schellinck, P. and Wisse, E. (1986). *Histochemistry*, **84**, 355.
8. Thornell, L.-E., Butler-Browne, G.S., Carlsson, E., Eppenberger, H.M., Fürst, D.O., Grove, B.K., Holmbom, B. and Small, J.V. (1986). *Scan. Electron Microsc.*, **4**, 1407.
9. Pietschmann, S.M., Gelderblom, H.R. and Pauli, G. (1989). *Arch. Virol.*, **108**, 1.
10. North, A.J., Galazkiewicz, B., Byers, T.J., Glenney, J.R. Jr. and Small, J.V. (1993). *J. Cell Biol.*, **120**, 1159.
11. Kudo, A., Fukushima, H., Kawakami, H., Matsuda, M., Goya, T. and Hirano, H. (1996). *J. Histochem. Cytochem.*, **44**, 615.
12. Tokuyasu, K.T., Slot, J.W. and Singer, S.J. (1978). In *Proceedings of the Ninth International Congress on Electron Microscopy*, Vol. 2, (ed. J. M. Sturgess). p. 164. Microscopical Society of Toronto, Canada.
13. Semper, A.E., Fitzsimons, R.B. and Shotton, D.M. (1988). *J. Neurol. Sci.*, **83**, 93.
14. Griffiths, G. (1993). *Fine Structure Immunocytochemistry*, p. 459. Springer-Verlag, Heidelberg.
15. Tokuyasu, K.T. and Singer, S.J. (1976). *J. Cell Biol.*, **71**, 891.
16. Malecki, M. and Small, J.V. (1987). *Protoplasma*, **139**, 160.
17. Gorbsky, G. and Borisy, G.G. (1986). *J. Histochem. Cytochem.*, **34**, 177.
18. Small, J.V., Fürst, D.O. and De Mey, J. (1986). *J. Cell Biol.*, **102**, 210.
19. North, A.J., Gimona, M., Lando, Z. and Small, J.V. (1994). *J. Cell Sci.*, **107**, 455.
20. North, A.J., Chidgey, M.A.J., Clarke, J.P., Bardsley, W.G. and Garrod, D.R. (1996). *Proc. Natl Acad. Sci. USA*, **93**, 7701.
21. Gosselin, E.J., Cate, C.C., Pettengill, O.S. and Sorenson, G.D. (1986). *Am. J. Anat.*, **175**, 135.
22. Bullock, G.R. (1984). *J. Microsc.*, **133**, 1.

23. Karnovsky, M.J. (1965). *J. Cell Biol.,* **27**, 137a.
24. Chiovetti, R., McGuffee, L.J., Little, S.A., Wheeler-Clark, E. and Brass-Dale, J. (1987). *J. Electron Microsc. Tech.,* **5**, 1.
25. Osborn, M. and Isenberg, S. (1994). In *Cell Biology: A Laboratory Handbook* (ed. J.E. Celis), Vol. 2, p. 361. Academic Press, San Diego, CA.
26. Barthel, L.K. and Raymond, P.A. (1990). *J. Histochem. Cytochem.,* **38**, 1383.
27. Ordonez, N.G., Manning, J.T. and Brooks, T.E. (1988). *Am. J. Surg. Pathol.,* **12**, 121.
28. Shi, S.-R., Key, M.E. and Kalra, K.L. (1991). *J. Histochem. Cytochem.,* **39**, 741.
29. Hazelbag, H.M., Van den Broek, L.J.C.M., Van Dorst, E.B.L., Offerhaus, G.J.A., Fleuren, G.J. and Hogendoorn, P.C.W. (1995). *J. Histochem. Cytochem.,* **43**, 429.
30. Bourguignon, L.Y.W., Tokuyasu, K.T. and Singer, S.J. (1978). *J. Cell. Physiol.,* **95**, 239.
31. Geuze, H.J. and Slot, J.W. (1980). *Eur. J. Cell Biol.,* **21**, 93.
32. Jensen, H.L. and Norrild, B. (1998). *J. Histochem. Cytochem.,* **46**, 487.
33. Tokuyasu, K.T. (1983). *J. Histochem. Cytochem.,* **31**, 164.
34. Glauert, A.M. (1975). *Fixation, Dehydration and Embedding of Biological Specimens: Practical Methods in Electron Microscopy,* Vol. 3, part I (Laboratory Edition). (ed. A. M. Glauert). North Holland, Amsterdam.
35. Mosedale, D.E., Metcalfe, J.C. and Grainger, D.J. (1996). *J. Histochem. Cytochem.,* **44**, 1043.
36. Bendayan, M., Nanci, A. and Kan, F.W.K. (1987). *J. Histochem. Cytochem.,* 35, 983.
37. Stierhof, Y.-D. and Schwarz, H. (1989). In *The Science of Biological Specimen Preparation for Microscopy and Microanalysis.* Proceedings of the Seventh Pfefferkorn Conference (ed. R.M. Albrecht and R.L. Ornberg), p.27. SEM Inc., AMF O'Hare, Chicago, MI.
38. Tokuyasu, K.T. (1986). *J. Microsc.,* **143**, 139.
39. Tokuyasu, K.T. (1990). *Histochem. J.,* **21**, 163.
40. Griffiths, G., McDowall, A., Back, R. and Dubochet, J. (1984). *J. Ultrastruct. Res.,* **89**, 65.
41. Liou, W., Geuze, H. and Slot, J,W. (1996). *Histochem. Cell Biol.,* **106**, 41.
42. Hunyady, B., Krempels, K., Harta, G. and Mezey, E. (1996). *J. Histochem. Cytochem.,* **44**, 1353.
43. Wehrens, X.H.T., Mies, B., Gimona, M., Ramaekers, F.C.S., Van Eys, G.J.J.M. and Small, J.V. (1997). *FEBS Lett.,* **405**, 315.
44. Johnson, G.D. (1989). In *Antibodies: A Practical Approach* (ed. D. Catt), p. 179. IRL Press, Oxford.
45. Osborn, M. (1981). *Tech. Cell Physiol.,* **107**, 1.
46. Lewis Carl, S.A., Gillete-Ferguson, I. and Ferguson, D.G. (1993). *J. Histochem. Cytochem.,* **41**, 1273.
47. Hierck, B.P., Iperen, L.V., Gittenberger-De Groot, A.C. and Poelmann, R.E. (1984). *J. Histochem. Cytochem.,* **42**, 1499.
48. Lan, H.Y., Mu, W., Nikolic-Paterson, D.J. and Atkins, R.C. (1995). *J. Histochem. Cytochem.,* **43**, 97.
49. Uchihara, T., Kondo, H., Akiyama, H. and Ikeda, K. (1995). *J. Histochem. Cytochem.,* **43**, 103.
50. Falck, B. and Owman, J. (1965). *Acta Univ. Lund,* sect. **II**, 7.
51. Van de Lest, C.H.A., Versteeg, E.M.M., Veerkamp, J.H. and Van Kuppevelt, T.H. (1995). *J. Histochem. Cytochem.,* **43**, 727.

52. Marriott, G., Clegg, R.M., Arndt Jovin, D.J. and Jovin, T.M. (1991). *Biophys. J.*, **60**, 1374.
53. Giloh, H. and Sedat, J.W. (1982). *Science*, **217**, 1252.
54. Valnes, K. and Brandtzaeg, P. (1985). *J. Histochem. Cytochem.*, **33**, 755.
55. Longin, A., Souchier, C., Ffrench, M. and Bryon, P.-A. (1993). *J. Histochem. Cytochem.*, **41**, 1833.
56. Berrios, M. and Colflesh, D.E. (1995). *Biotech. Histochem.*, **70**, 40.
57. Germroth, P.G., Gourdie, R.G. and Thompson, R.P. (1995). *Microsc. Res. Tech.*, **30**, 513.

3

Simultaneous *in situ* detection of DNA and proteins

KLAUS ERSFELD and ELISA M. STONE

1. Introduction

The development of fluorescent *in situ* hybridization (FISH) in combination with digital imaging technology has contributed greatly to our current view on how the nucleus of eukaryotic cells is organized. It has been shown that chromatin is not randomly distributed but is confined to defined sub-regions of the nucleus (1–4). There is mounting evidence that the non-random chromatin distribution is caused by anchoring of chromosomal sub-regions to elements of the nucleoskeleton, such as the nuclear lamina. The compartmentalization of genetic material and its interaction with specific proteins has important functional implications for various processes such as replication (5), transcription (6) or gene silencing (7–10) and chromosome partitioning (11–15) (*Figure 1A–G*). In many cases the experimental approaches used to demonstrate the interaction of proteins and DNA in the three-dimensional framework of the nucleus includes the simultaneous detection of proteins by immunofluorescence and chromosomal regions by FISH. In this chapter we provide protocols for these techniques that have been adapted from those previously described for other organisms, and that are therefore generally applicable. Specifically, we have used such methods extensively for our work on trypanosome nuclear organization (15). Because a significant number of studies on nuclear architecture, particularly relating to gene silencing mechanisms, have been done in yeast, we also include a method for preparing yeast cells for these procedures.

2. Preparation of labelled DNA probes

In this paragraph we describe several techniques for labelling DNA with non-radioactively modified nucleotides. Although a range of different reporter nucleotides are now commercially available, digoxigenin- and biotin-labelled derivatives of dUTP are frequently used and several companies offer kits to perform the labelling procedures described below. For most applications, labelling by nick translation is the method of choice because a wide range of substrate DNAs can be used, including long genomic clones (e.g. cosmids, P1 plasmids, bacterial or yeast artificial chromosomes (BACs and YACs), and pools of plasmids) which will cover enough area on a chromosome to enable detection. Labelling by PCR is useful if highly repetitive sequence elements are the target, such as tandemly repeated genes. In addition, other PCR-based techniques (e.g. degenerate oligonucleotide primed PCR) have been developed to label small quantities of whole chromosomes isolated by pulsed-field gel electrophoresis or flow sorting (16,17). Short repetitive sequences, in particular telomeric repeats, can be visualized using end-labelled oligonucleotides. In many cases the simultaneous detection of proteins and chromosomal DNA is used to find out whether a particular protein co-localizes with a defined chromosomal element. This requires a high degree of structural preservation, thereby excluding extraction methods commonly used in other FISH applications such as proteinase digestions. As a consequence, the sensitivity of the FISH procedure is comparatively low. In our experience with FISH in trypanosomes it is necessary to have 15–20 kb of continuous target DNA on a given chromosome to achieve visualization. This has to be kept in mind when selecting DNA probes for labelling.

2.1 Labelling of DNA by nick translation

When using nick translation for *in situ* hybridization applications (18,19), it is possible to control the final length of the labelled product by titrating the amount of DNase I. For FISH the optimal probe length is 100–500 bp. Longer probes may result in increased background and shorter probes lead to a decrease in signal intensity. The source of DNA for labelling can be a cloned DNA fragment in a plasmid vector, genomic DNA or PCR products. In this protocol we describe the incorporation of digoxigenin- and biotin-modified nucleotides into DNA. Commercial nick translation kits that have been optimized for FISH applications are also available from Roche Diagnostics and Gibco-BRL.

Protocol 1. Nick translation for FISH probes

Reagents

- 10× nick translation buffer: 0.5 M Tris–Cl pH 7.8, 50 mM $MgCl_2$, 0.5 mg/ml bovine serum albumin (BSA) (store at –20°C)
- 0.1 M dithiothreitol (DTT) in double-distilled H_2O (store at –20°C)
- 10× nucleotide stock solution: 0.5 mM dATP, 0.5 mM dCTP, 0.5 mM dGTP, 0.34 mM dTTP, 0.16 mM digoxigenin-11-dUTP (Roche Diagnostics) or biotin-11-dUTP (Roche Diagnostics or Gibco-BRL) in double-distilled H_2O (store at –20°C)

- DNase I stock solution: dissolve lyophilized pancreatic DNase I (Roche Diagnostics) in 0.15 M NaCl, 50% (v/v) glycerol at 1 mg/ml and store at –20°C
- DNA polymerase I (5 units/µl) (Roche Diagnostics) (store at –20°C)
- 0.5 M EDTA in double-distilled H_2O
- TE: 10 mM Tris–Cl, 1 mM EDTA pH 7.5, prepared according to standard methods (20)
- 3 M sodium acetate pH 5.8

- SSC: 150 mM NaCl, 150 mM sodium citrate pH 7.0, prepared according to standard methods (20)
- 10 mg/ml herring sperm DNA: dissolve DNA (e.g. Sigma) in TE to give a concentration of 10 mg/ml. Sonicate this solution and check fragments on an agarose gel. The size range should be 100–1000 bp. (store at –20°C)
- 10 mg/ml yeast tRNA (e.g. Sigma or Roche Diagnostics) in TE (store at –20°C)
- 2× hybridization buffer without formamide: 4× SSC, 20% (w/v) dextran sulfate, 100 mM sodium phosphate buffer pH 7.0 (store at –20°C)

- Ethanol, 100% and 70% in double-distilled H_2O
- Formamide: commercially available formamide must be deionized before use. In a 1.5 ml reaction tube, mix 1 g of mixed bed resin ion exchanger beads (e.g. TMD-8, Sigma) with 1 ml formamide. Leave on ice for at least 10 min. The formamide is now ready for use. Make fresh every day.
- Hybridization buffer: mix 1 vol. of 2× hybridization buffer with 1 vol. of deionized formamide. Final concentrations are: 50% formamide, 2× SSC, 10% dextran sulfate, 50 sodium phosphate pH 7.0. Keep on ice and use the same day.

Method

1. Immediately before use, dilute DNase I stock solution 1/1000 in ice-cold double-distilled H_2O and keep on ice.

2. For the labelling reaction combine in a reaction tube on ice, adding double-distilled H_2O as the first component of the reaction (final reaction volume 50 μl):

   - 10× nick translation buffer, 5 μl
   - 0.1 M DTT, 5 μl
   - 10× nucleotide stock solution, 5 μl
   - template DNA, 1 μg
   - DNA polymerase I (5 units/μl), 4 μl
   - diluted DNase I (1 μg/ml), 5 μl

3. Incubate for 2 h at 15°C in a water bath (a 2 L volume of water keeps the temperature constant for that period of time).

4. Stop the reaction by adding 1 μl of 0.5 M EDTA and heating at 70°C for 10 min. Store at –20°C.

5. Check 5 μl of the reaction by gel electrophoresis on a 1.3% (w/v) agarose gel and stain with ethidium bromide. There should be a smear of DNA between 100 and 500 bp. If the DNA is considerably smaller or larger it will be necessary to repeat the labelling reaction with decreased or increased amounts of DNase I. A labelling reaction resulting in larger products can still be used for DNA or RNA blotting applications. Due to the high costs of modified nucleotides, an initial titration series of variable amounts of DNase I with unmodified nucleotides only is recommended.

6. It is not necessary to purify the probe, but because the FISH procedure requires a different buffer system and the inclusion of carrier nucleic acids, it is convenient to co-precipitate the probe with the carrier material and dissolve the pellet in hybridization buffer.

Protocol 1. *Continued*

7. To precipitate the amount of probe necessary for one slide, combine the following:

   - terminated labelling reaction, 1 μl (~20 ng)
   - herring sperm DNA, 1 μl
   - yeast tRNA, 1 μl
   - double-distilled H_2O, 6 μl
   - 3 M sodium acetate pH 5.8, 1 μl
   - 100% ethanol (at room temperature), 25 μl

8. Spin at full speed in a bench-top centrifuge for 20 min, remove supernatant, wash pellet with 50 μl 70% ethanol, spin for 5 min, remove supernatant and let pellet dry at room temperature for 10–20 min. Dissolve the pellet in 25 μl of hybridization buffer and keep on ice. If the probe is not used on the same day, it can be stored at –80°C for at least a week.

2.2 Labelling of DNA by PCR

In this protocol we describe the incorporation of digoxigenin- or biotin-modified nucleotides into DNA using PCR. The main advantages of labelling DNA by PCR are the large amounts of probe generated during the reaction in contrast to a smaller amount of net synthesis during nick translation and the precise control of probe length as determined by the choice of oligonucleotide primers. Similar to nick translation, probe length should not exceed 500 bp. If longer probes are generated, they can be fragmented with specific restriction enzymes, provided the DNA sequence is known, or with frequent cutters, such as *Alu*I. Our preferred substrate for a PCR-based labelling reaction is a purified PCR product. It is important to remove free nucleotides from the first PCR reaction, because the presence of dTTP will decrease the labelling efficiency. A number of different kits, such as the Qiagen PCR purification kit, are suitable.

Protocol 2. Labelling by PCR for FISH probes

Reagents

- 10× PCR buffer: as supplied with *Taq* DNA polymerase by the manufacturer
- *Taq* DNA polymerase, 5 units/μl (e.g. from Roche Diagnostics or Promega)
- dNTP mix, without dTTP: 5 mM dATP, 5 mM dCTP, 5 mM dGTP in double-distilled H_2O
- 5 mM dTTP solution in double-distilled H_2O
- 1 mM solutions of either digoxigenin-11-dUTP (Roche Diagnostics) or biotin-11-dUTP in ddH$_2$O (Roche Diagnostics or Gibco-BRL)
- Oligonucleotide primers at a concentration of 10–50 pmol/μl

---

Method

1. Combine in a 0.5 ml reaction tube:

   - 10× PCR buffer, 5 μl
   - primer A (0.1–0.5 μM final concentration), 1 μl
   - primer B (0.1–0.5 μM final concentration), 1 μl
   - dNTP mix (without dTTP), 2 μl
   - dTTP, 1.3 μl
   - digoxigenin-dUTP or biotin-dUTP, 3.5 μl
   - template DNA, 10 ng
   - *Taq* DNA polymerase, 1 μl
   - H$_2$O, add to 50 μl total volume.

2. Spin briefly and overlay with 50 μl of mineral oil.

3. Carry out PCR under the following cycling conditions:

   (a) 30 cycles of 1 min at 95°C, 1 min at 50°C, 2 min at 72°C;

   (b) final extension: 10 min at 72°C.

4. Remove the aqueous phase and store at –20°C. Analyse 3 μl by agarose gel electrophoresis. Run the labelled fragment alongside the unlabelled template fragment. There should be a considerable shift towards a higher molecular weight of the labelled fragment, indicating a successful labelling.

5. Purify the reaction using a PCR purification kit (Qiagen). Elute with 50 μl TE, pH 8.0. For many FISH applications, 1 μl of a 1/10 dilution of this probe per slide is sufficient. Co-precipitate and process the desired amount of labelled DNA together with carrier DNA and RNA as described in *Protocol 1*.

2.3 3′-Labelling of oligonucleotides for FISH probes

To detect DNA targets consisting of long, repetitive sequences, such as telomeres, oligonucleotides can be used as probes. Their main advantage is that they are single stranded and have very fast hybridization kinetics. Typical hybridization times to detect telomeric sequences are 3–6 h. In *Protocol 3* we describe the 3′-labelling of oligonucleotides with digoxigenin-11-ddUTP resulting in the addition of a single modified nucleotide to the oligonucleotide. The protocol has been adapted from the *In Situ Hybridization Application Manual* published by Roche Diagnostics.

Protocol 3.   3′-Labelling of oligonucleotides for FISH probes

Reagents

- 5× reaction buffer: 1 M potassium caco-dylate,[a] 125 mM Tris–Cl, 1.25 mg/ml BSA pH 6.6
- 1 mM digoxigenin-11-ddUTP (Roche Diagnostics)

- 25 mM aqueous $CoCl_2$
- Terminal transferase (50 units/μl) (Roche Diagnostics)
- Oligonucleotide (~100 pmol/μl)
- 0.5 M EDTA pH 8.0

Method

1. Combine in a reaction tube on ice, adding 9 μl H_2O first for a final volume of 20 μl:
   - 5× reaction buffer, 4 μl
   - $CoCl_2$, 4 μl
   - oligonucleotide, 1 μl
   - digoxigenin-11-ddUTP, 1 μl
   - terminal transferase, 1 μl
2. Incubate at 37 °C for 20 min.
3. Stop the reaction by adding 1 μl of 0.5 M EDTA.
4. Purification is not necessary; store at –20 °C.
5. The optimal amount for FISH must be determined for each oligonucleotide, but 0.5 μl of the reaction per slide often is a good starting point. Co-precipitate and process the desired amount of labelled oligonucleotide together with carrier DNA and RNA as described in *Protocol 1*.

[a] Potassium cacodylate is toxic. Handle with care when preparing the solution. Usually this buffer is provided with the enzyme by the manufacturer.

---

3. Preparation of cells

The aim of the combined FISH/immunofluorescence technique is to correlate the localization of chromosomal or other DNA elements with the localization of proteins. Therefore, the conservation of the structural integrity is of paramount importance. One has to find a balance between sensitivity, particularly as far as FISH is concerned, and structural preservation. This protocol tries to keep the extraction of cellular components at a minimum in order to avoid disrupting cellular and nuclear structure, and at the same time minimizing the compromise in FISH detection sensitivity. We fix cells with formaldehyde and subsequently permeabilize membranes with a non-ionic detergent to allow access of antibodies to intranuclear structures. We have used this technique extensively to study chromosome dynamics in *Trypanosoma brucei*.

Protocol 4. Coating slides with organosilane

To improve adherence of cells, glass microscope slides are coated with organosilane. Alternatively, poly-L-lysine coating techniques can be used (Chapter 1, *Protocol 1*), although they sometimes lead to increased background. The following protocol has been adapted from Nuovo (21).

Equipment and reagents
- 3-Aminopropyltriethoxy-silane (Sigma)
- High-grade acetone
- 2 M HCl
- Standard microscope slides
- Coplin jars

Method
1. Put slides in a Coplin jar and wash in 2 M HCl for 5 min.
2. Rinse in double-distilled H_2O for 1 min.
3. Wash in acetone for 1 min and air dry.
4. Coat slides in a 2% (v/v) silane solution in acetone for 1 min.
5. Wash slides in acetone for 1 min and air dry.
6. Slides can be kept in a dust-free box for 3–4 weeks at room temperature but deteriorate over longer periods.

Protocol 5. Cell fixation and permeabilization

Reagents
- PBS: 140 mM NaCl, 3 mM KCl, 10 mM Na_2HPO_4, 1.4 mM KH_2PO_4, pH 7.4
- Permeabilization solution: 0.1% Nonidet P-40 (v/v) (ICN or Sigma) in PBS
- Fixation solution: 3.6% (v/v) formaldehyde (EM-grade, TAAB), 5% acetic acid (v/v) in PBS

Method[a]
1. Mark an area of $\sim$1 cm $\times$ 1 cm on the microscope slide from underneath with a diamond pen.
2. For trypanosome cells grown in suspension: pellet cells by low-speed centrifugation and resuspend pellet in PBS. Either put a 50–100 μl drop of the cell suspension on the marked area of a silanized slide and let cells settle for 5 min in a humid chamber or spin the cells onto the slide using a cytospin centrifuge (Shandon).
3. For cells grown on coverslips: dip the coverslip briefly in PBS and continue with the fixation immediately.
4. Fix the cells in fixation solution for 15 min at room temperature in a

Protocol 5. *Continued*

humid chamber by overlaying the marked area of the slide or the coveslip with 100–500 μl of fixation solution.

5. Wash the slides or coverslips for 2 × 5 min in PBS using a Coplin jar.

6. Permeabilize the cells by overlaying the slide or coverslip with 100–500 μl of permeabilization solution.

7. Wash the slides or coverslips for 5 min in PBS using a Coplin jar.

8. The cells are now ready for incubation with the primary antibody.

[a] Do not let the cells dry at any stage of this procedure.

4. Combined immunofluorescence and *in situ* hybridization

In *Protocol 6* we describe the detection of trypanosome nuclear antigens with specific primary antibodies and secondary fluorochrome-conjugated antibodies, and the subsequent detection of specific DNA targets using either digoxigenin-conjugated or biotinylated probes. Although it is possible to do the antigen detection step after the FISH procedure, immunofluorescent detection is of higher quality if done prior to FISH because cellular structures are better preserved at that stage. After immunodetection the antigen–antibody complex is stabilized by cross-linking with formaldehyde. Because antigen and DNA detection both involve the use of antibodies, care should be taken in the choice of antibody combinations to avoid cross-reactions between the different detection steps. Obviously, the fluorochromes used in the two detection steps must be compatible. Examples of combined FISH and immunofluorescence using *T. brucei* as an experimental model are shown in *Figure 1*, panels A–G.

Protocol 6. Combined antigen and DNA detection by immunofluorescence and *in situ* hybridization

Equipment and reagents

- Formaldehyde (see *Protocol 4*)
- PBS (see *Protocol 4*)
- 20× SSC (see *Protocol 1*)
- 70%, 90% and 100% (v/v) ethanol at –20°C
- RNase I, 10 mg/ml in double-distilled H$_2$O (boil for 10 min and allow to cool slowly to room temperature; store at –20°C)
- Hybridization buffer (see *Protocol 1*)
- Antibody dilution buffer: 100 mM maleic acid pH 7.5, 150 mM NaCl, 1% (w/v) blocking reagent (Roche Diagnostics)

- Formamide (molecular biology grade; e.g. from Sigma or Oncor-Appligene)
- Antibody specific for your favourite molecule
- Secondary, fluorochrome-conjugated antibody (e.g. from Sigma, Jackson Laboratories, or DAKO)
- Anti-digoxigenin Fab fragments made in sheep (1 mg/ml) (Roche Diagnostics)
- FITC-conjugated rabbit-anti-sheep antibodies (1.5 mg/ml) (Vector Immunoresearch)

- Cy3-conjugated anti-biotin reagents, e.g. Cy3–ExtrAvidin (Sigma) or Cy3–streptavidin (Jackson ImmunoResearch)
- PBST: 0.05% (v/v) Tween 20 in PBS
- Mounting medium: Vectashield (Vector Immunoresearch) or any of the anti-fade mounting media described in Chapter 1, including 100 ng/ml 4,6-diamidine-2-phenylindole dihydrochloride (DAPI)
- Coverslips
- Water baths set at hybridization temperature (34–38°C) and 50°C
- Rubber cement (available from art and craft shops as 'cow gum' rubber). A convenient alternative is EasySeal plastic coverslips (Hybaid). Using a self-adhesive plastic frame, the EasySeal coverslips can be glued onto the slides and provide a small volume for the hybridization solution. After hybridization they can easily be pulled off the glass slide.
- Coplin jars
- Thermal cycler with *in situ* hybridization block (e.g. Hybaid)[a]

Method

11. After the fixation of the cells is completed (*Protocol 4*) place 50 µl of the primary antibody on the slide. Hybridoma supernatants are often used undiluted. If dilution of the antibodies is necessary (e.g. for serum or purified monoclonals) use antibody dilution buffer for this purpose. To prevent evaporation, cover the antibody solution on the slide with a small square of Parafilm. Incubate for 1 h at 37°C. (If cross-hybridization between mRNA and the DNA probe might be a problem in the subsequent FISH procedure, include 10 µg/ml RNase I in the first antibody solution to digest the RNA.)

12. Wash the slides for 3 × 5 min in PBST in a Coplin jar.

13. Incubate cells with the secondary, fluorochrome-conjugated antibody (diluted in antibody dilution buffer) for 1 h at 37°C.

14. Wash as above.

15. Fix the antigen–antibody complex by overlaying the cells on the slide with 100 µl of 3.6% (w/v) formaldehyde in PBS for 15 min at room temperature.

16. Wash slides for 3 × 5 min in PBS.

17. Equilibrate the cells with 50 µl of hybridization buffer for 30 min at room temperature. To prevent evaporation, cover the solution with a small square of Parafilm.[a]

18. Remove as much of the hybridization buffer as possible and add 25 µl of hybridization solution containing the labelled DNA probe. Cover the marked area with a coverslip and seal with 'cow gum' rubber cement. This is conveniently done using a 2 ml syringe. If you use EasySeal coverslips, follow the instructions of the manufacturer to seal the slide.

19. Place slide on the *in situ* block of the thermal cycle and start the following program:
    - 5 min at 85°C;
    - 16 h at 38°C for DNA probes (for oligonucleotides use 34°C).

Protocol 6. *Continued*

10. After hybridization, remove rubber seal (leave coverslip in place) and wash slide in 50% (v/v) formamide, 2× SSC, pre-warmed to hybridization temperature, for 30 min. Use Coplin jars in a water bath set to the appropriate temperatures. The coverslip will slide off after a few minutes. If you use EasySeal coverslips, pull them off the slide before the washing step. Up to five slides can be processed in one jar.

11. Wash slides in 2× SSC for 10 min at 50°C, for 2 × 15 min in 0.2× SSC at 50°C (0.5× SSC for oligonucleotide probes) and in 4× SSC for 10 min at room temperature.

12. Incubate slide with 50 μl of the anti-digoxigenin Fab fragment (0.3 μg/ml in antibody dilution buffer) for 1 h at 37°C in a humid chamber. If a biotinylated probe has been used, incubate with the appropriate avidin or streptavidin conjugate and proceed to step 15.

13. Wash in PBST for 3 × 5 min at room temperature

14. Incubate with anti-digoxigenin conjugate, diluted to 10 μg/ml in antibody dilution buffer, for 1 h at 37°C in a humid chamber.

15. Wash in PBST for 3 × 5 min at room temperature.

16. Wash slides for 10 sec in double-distilled H_2O. After the wash, remove as much liquid from the slide as possible.

17. Add 10 μl of embedding medium on the marked area and cover with a glass coverslip. Seal edges with nail varnish.

---

[a] If an *in situ* thermal cycler is not available, the target and probe must be denatured separately. Heat the probe, dissolved in hybridization buffer, at 90°C for 5 min and chill on ice water. Denature the cellular DNA by immersing the slide in a solution of 70% (v/v) formamide in double-distilled H_2O, pre-warmed to 75°C, for 5 min. Use a Coplin jar in a heated water bath. Make sure the temperature *inside* the jar is at 75°C and do not use more than two slides per jar. Then dehydrate the slides through 70%, 90% and 100% (v/v) ethanol at –20°C for 3 min each and air dry them. Add the denatured probe, seal as described in step 8 and hybridize at 34–38°C for 12–16 h in a conventional incubator. Proceed with step 10.

5. Applications of combined immunofluorescence and FISH in yeast

5.1 Co-localizing chromosomal proteins and chromosomal domains

In the budding yeast *Saccharomyces cerevisiae*, the transcriptional silencing proteins Rap1p, Sir3p and Sir4p were localized to discrete foci by indirect immunofluorescence (7,22). These foci represent the ends of the chromosomes—the telomeres—as demonstrated with the use of a telomere-specific

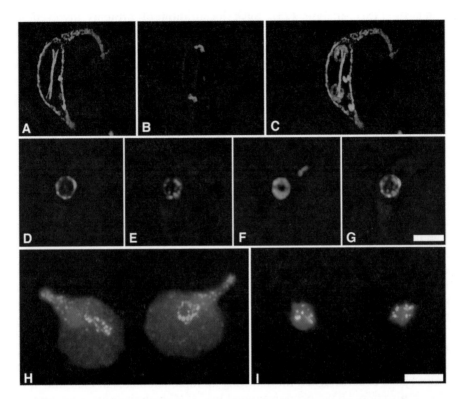

Figure 1. (A–C) Visualization of microtubules including the mitotic spindle, a chromosome sub-population (minichromosomes) and total DNA in a mitotic *Trypanosoma brucei* cell. Microtubules were labelled with an anti-β-tubulin monoclonal antibody and visualized with a secondary, FITC-conjugated antibody (panel A, green). Minichromosomes were detected by FISH using a PCR-generated, biotinylated DNA probe and a Cy3–ExtrAvidin conjugate (panel B, red). Cells were stained with DAPI for total DNA (blue in merged image, panel C). Reprinted with permission from ref. 15. Copyright (1997) American Association of the Advancement of Science. (D and E) Localization of the nuclear envelope and telomeres in an interphase *T. brucei* cell. A protein of the nuclear envelope was visualized with a monoclonal antibody and a fluorochrome-conjugated secondary antibody (panel D, green). Telomeres were detected with a digoxigenin end-labelled oligonucleotide, an anti-digoxigenin secondary antibody and a Cy3-conjugated tertiary antibody (panel E, red). Total DNA was stained with DAPI (panel F, blue). Panel G shows the merged image. (H and I) Yeast cells respond to mating pheromone by undergoing a characteristic rearrangement of the cytoskeleton and nuclear reorganization (30). Panel H, immunofluorescent actin-staining visualizes the mating projection tip (red), and detection of ribosomal DNA by FISH detects an unusual nucleolar morphology (green). Panel I, Immunofluorescent tubulin staining detects the mitotic spindle (red), and clustered telomeres are detected by FISH using a probe against the repetitive Y{pri} telomeric sequences (green). DAPI stains the total chromosomal DNA (blue). Scale bars represent 2 μm.

probe by FISH analysis (23). Such localization of proteins involved in chromatin structure to specific chromosomal loci in budding yeast has been achieved by combined immunofluorescence and FISH techniques, for which methods have been described (24). An analogous technique has been used to co-localize the chromodomain protein Swi6p with centromeres, telomeres and the silent mating-type loci in the fission yeast *Schizosaccharomyces pombe* (8).

5.2 Co-localizing cytoskeletal proteins and chromosomal domains

Combined immunofluorescence and FISH techniques in *S. pombe* were developed to map the position of chromosomal loci such as rDNA, centromeres and telomeres relative to microtubules and the yeast centrosome, the spindle pole body (13,25). In *S. cerevisiae*, components of the cytoskeleton have been co-localized effectively with chromosomal loci using green fluorescent protein (GFP) tagging techniques (26). The advantage of this approach is that live cells can be examined rapidly, avoiding possible distortion of cell shape during the rigorous experimental conditions required for FISH. It is limited, however, in the number of loci that can be examined simultaneously. Thus far, only one or two centromeres or telomeres have been tagged with the Lac operator marker; this does not allow for observation of the behaviour of the entire population of centromeres or telomeres. Alternatively, FISH probes have been designed for detecting multiple telomeres at the same time (23), centromeres have been examined by FISH analysis using a mixture of different centromere-specific cosmids (27,28), and ribosomal DNA (rDNA) FISH has been reported for detecting the nucleolus (29); however, simultaneous immunofluorescence to detect cytoskeletal elements was not performed in these studies.

In an effort to study nuclear organization in the context of cytoskeletal rearrangement in budding yeast cells undergoing the mating pheromone response (30), previously described methods (24) were modified. A fixation protocol, presented below, was developed for optimal detection of chromosomal loci by *in situ* hybridization or by immunofluorescent staining of chromosomal proteins, and co-localization of cytoskeletal elements (30). In this protocol, by fixing cells before the enzymatic digestion of the cell wall that is required for penetration of antibodies and FISH probes, both cell morphology and the integrity of the nucleus are preserved. In *Figure 1* panel I we show an example of mating pheromone-responding cells in which telomeres are detected by FISH and microtubules spanning the nuclei can be seen by immunofluorescence. Using different probes, the rDNA loci are detected by FISH and actin staining identifies the mating projection tip by immunofluorescence (*Figure 1H*).

Protocol 7. Fixing cells for combined immunofluorescence and FISH in budding yeast

Equipment and reagents

- Prepare a 20% (w/v) paraformaldehyde solution in a loosely capped bottle (**Caution:** hazardous[a]). Add NaOH from a stock solution to a final concentration of 1 mM and heat carefully in a microwave on low setting. Remove the bottle just as the contents begin to boil, and tighten the cap, swirling to mix, opening the bottle to vent in a fume hood. Repeat this step several times until the paraformaldehyde has dissolved. Place on ice.
- 0.1 M EDTA pH 8.0, 10 mM DTT

- YPD medium (prepared as described in ref. 31), with and without 1.2 M sorbitol
- Yeast lytic enzyme isolated from *Arthrobacter luteus*, 70 000 U/g (ICN)
- 10-well slides (HTC Supercured, Cel-line Associates, Inc.)
- 100% methanol at –20 °C
- 100% acetone at –20 °C
- Blocking buffer: PBS containing 0.1% (v/v) Triton X-100 and 1% (w/v) ovalbumin

Method

1. Grow yeast in standard rich or selective growth medium (31) to a cell density of 1–2 × 10^7 cells/ml, corresponding to mid-log phase of growth. A 50 ml culture is recommended to ensure that there are enough cells to work with conveniently at the end of the procedure.

2. Fix cells by adding paraformaldehyde solution to the growing yeast culture to a final concentration of 3.3% (v/v). Shake for 10 min[b] at 30 °C.

3. Harvest cells at 1200g for 5 min at room temperature in pre-weighed 50 ml conical plastic tubes. Wash twice in 0.5 vol. of YPD medium. Weigh the cell pellet.

4. Resuspend cells at 1 ml/0.1 g cells, in 0.1 M EDTA pH 8.0 and 10 mM DTT. Incubate at 30 °C for 10 min with gentle agitation. Collect cells by centrifugation at 800g for 5 min.

5. Carefully resuspend the pellet, 1 ml/0.1 g cells, in YPD/1.2 M sorbitol (first resuspending in 1 ml with a P1000 pipette, then adding the remaining volume). To remove the cell wall, add yeast lytic enzyme to 0.6 mg/ml. Note that the enzyme concentration is three times that used in previously described methods (24). Incubate at 30 °C in a 50 ml Erlenmeyer flask with gentle agitation, monitoring for spheroplast formation under the microscope every 5 min (for approximately 25 min or more, up to 1 h, depending on the strain). As digestion of the cell wall takes place, cells first begin to darken and appear grainy, and then become transparent. Incubation should be stopped as soon as the first signs of cell lysis occur.

6. Dilute with 5 vols of YPD, and harvest cells by centrifugation at 800g for 5 min. Aspirate supernatant carefully, as the pellet may be quite loose. Wash twice more by resuspending cell pellet gently as before in 1 ml YPD/1.2 M sorbitol then adding up to 4–5 vols YPD/1.2 M sorbitol.

Protocol 7. *Continued*

Resuspend in YPD *without* sorbitol (in order for cells to stick easily to slides) to appropriate spheroplast density (e.g. 2 ml for 0.3 g original cell pellet).

7. Drop spheroplasts into each well of a 10-well slide, wait for 1–2 min as cell suspension is held in well by surface tension, then remove liquid and air-dry the slide for 2 min. Save the remaining cells in YPD at 4°C for pre-clearing secondary antibodies. Place the slide in a –20°C Coplin jar methanol bath for 6 min, then in a –20°C acetone bath for 1 min. Air dry for 3 min. Check cell density under a microscope; a single layer of cells with many cells in each field is ideal. Rehydrate slide in blocking buffer, and continue immunofluorescence and FISH procedures as described in ref. 23.

[a] Paraformaldehyde is highly toxic. It is readily absorbed through the skin and is extremely destructive to skin, eyes, mucous membranes and upper respiritory tract. Paraformaldehyde powder and formaldehyde stock solutions should always be handled in a fume hood. Protective clothing and gloves should be worn when handling powder and solutions.
[b] This relatively short fixation time is useful for antibodies against the silent chromatin proteins Sir3p, Sir4p and Rap1p, for nucleolar proteins and for actin, microtubules and spindle pole body antigens. It may be necessary to perform a time course of fixation for other antibodies, although digestion of the cell wall may require longer incubation and a higher concentration of enzymes with longer fixation times.

6. Microscopy and image analysis

In contrast to techniques such as the visualization of cytoskeletal proteins or FISH on metaphase chromosome spreads, nuclear cytology has to deal with two major constraints:

(a) the protein of interest may not be abundant; and

(b) the DNA target has a limited size, and accessibility of the probes is restricted by necessarily gentle, structure-preserving, fixation and extraction methods.

Furthermore, judging whether a particular protein co-localizes with certain chromosomal elements, e.g. telomeres, or the ability to measure chromosome dynamics, requires a microscope system which is capable of working close to the theoretical limits of optical resolution. In addition to a high quality, well-maintained research microscope we highly recommend the application of digital imaging. The use of digital cameras, such as high-sensitivity and high-resolution slow-scan CCDs, in combination with computerized image processing helps to eliminate errors in aligning images generated with different filter sets. Furthermore, image enhancement software, such as contrast adjustments and deconvolution, facilitates the analysis of data. See Chapters 4–6 for further details about imaging technology.

For example, for the images in *Figure 1A–G*, we have used a Leica DMRXA microscope equipped with a Photometrics cooled CH250 series slow-scan CCD camera. The objective is an 100× oil-immersion Plan-Apochromat (Leica). Filter sets for the various fluorochromes are from Chroma. The camera is controlled using IP-Lab-Spectrum image analysis software with an acquisition module plug-in (Scanalytics, Inc.) on a Macintosh 8100/100 PowerPC with 128 megabytes of RAM and a 4 gigabyte hard disc. The computer is linked to an external 600 megabyte magneto-optical disc (MOD) drive and a CD writer. We recommend saving images initially on the computer hard drive and then copying the files onto MOD afterwards. In our experience MODs are not the most reliable long-term storage medium and therefore it is extremely important to have backup copies of your data, preferably on CD. New storage media, such as ZIP or JAZ drives, are likely to supersede MOD drives. Sometimes images are deblurred using a two-dimensional deconvolution program (Hazebuster, Vaytek, Inc.). Captured greyscale images are pseudocoloured, merged and annotated in Adobe Photoshop. For *Figure 1* panels H and I, the microscope was equipped with a SensiCam CCD camera (PCO), and images were captured and manipulated using the SlideBook software package (Intelligent Imaging Innovations). Because most scientific journals accept electronic image files, an expensive, photographic quality printer is not necessary. For most purposes a colour inkjet printer (e.g. Epson Stylus Colour 800) is adequate. There are a number of manufacturers of high quality research microscopes, cameras and imaging software which can be used in different combinations. Choosing the optimal equipment will depend on the special applications of the individual laboratory (see Chapters 4–6).

Acknowledgements

We would like to thank S. Gasser, K. Gull, J. Lowell and L. Pillus for critical comments on the manuscript. The work in our laboratories is funded by a Wellcome Trust Programme Grant (K.E.) and NIH grant GM54778 (E.S.).

References

1. Marshall, W. F., Straight, A., Marko, J. F., Swedlow, J. F., Dernburg, A., Belmont, A., Murray, A. W., Agard, D. A. and Sedat, J. W. (1997). *Curr. Biol.*, **7**, 930.
2. Lamond, A. I. and Earnshaw, W. C. (1998). *Science*, **280**, 547.
3. Marshall, W. F., Fung, J. C. and Sedat, J. W. (1997). *Curr. Opin. Genet. Dev.*, **7**, 259.
4. Cremer, T. *et al.* (1993). *Cold Spring Harbor Symp. Quant. Biol.*, **58**, 777.
5. Hozak, P., Hassan, A. B., Jackson, D. A. and Cook, P. R. (1993). *Cell*, **73**, 361.
6. Xing, Y., Johnson, C. V., Moen, P. T., McNeal, J. A. and Lawrence, J. B. (1995). *J. Cell Biol.*, **131**, 1635.

7. Palladino, F., Laroche, T., Gilson, E., Axelrod, A., Pillus, L. and Gasser, S. M. (1993). *Cell*, **75**, 543.
8. Ekwall, K., Javerzat, J.-P., Lorentz, A., Schmidt, H., Cranston, G. and Allshire, R. (1995). *Science*, **269**, 1429.
9. Dernburg, A. F., Broman, K. W., Fung, J. C., Marshall, W. F., Philips, J., Agard, D. A. and Sedat, J. W. (1996). *Cell*, **85**, 745.
10. Csink, A. K. and Henikoff, S. (1996). *Nature*, **381**, 529.
11. Chibana, H. and Tanaka, K. (1996). *Genes to Cells*, **1**, 727.
12. Chikashige, Y., Ding, D., Imai, Y., Yamamoto, M., Haraguchi, T. and Hiraoka, Y. (1997). *EMBO J.*, **16**, 193.
13. Funabiki, H., Hagan, I., Uzawa, S. and Yanagida, M. (1993). *J. Cell Biol.*, **121**, 961.
14. Bass, H. W., Marshall, W. F., Sedat, J. W., Agard, D. A. and Cande, W. Z. (1997). *J. Cell Biol.*, **137**, 5.
15. Ersfeld, K. and Gull, K. (1997). *Science*, **276**, 611.
16. Telenius, H., Carter, N. P., Bebb, C. E., Nordenskjold, M., Ponder, B. A. J. and Tunnacliffe, A. (1992). *Genomics*, **13**, 718.
17. Bailey, D. M. D., Carter, N. P., de Vos, D., Leversha, M. A., Perryman, M. T. and Ferguson-Smith, M. A. (1993). *Nucleic Acids Res.*, **21**, 5117.
18. Rigby, P. W. J., Diekmann, M., Rhodes, C. and Berg, P. (1977). *J. Mol. Biol.*, **113**, 237.
19. Langer, P. R., Waldrop, A. A. and Ward, D. C. (1981). *Proc. Natl Acad. Sci. USA*, **78**, 6633.
20. Sambrook, J., Fritsch, E. F. and Maniatis, T. (1989). *Molecular Cloning. A Laboratory Manual*, 2nd edn. Cold Spring Harbor Laboratory Press, Cold Spring Harbor, NY.
21. Nuovo, G. R. (1994). *PCR In Situ Hybridization*, 2nd edn, p. 103. Raven Press, New York.
22. Klein, F., Laroche, T., Cardenas, M. E., Hofmann, J. F.-X., Schweizer, D. and Gasser, S. M. (1992). *J. Cell Biol.*, **117**, 935.
23. Gotta, M., Laroche, T., Formenton, A., Maillet, L., Scherthan, H. and Gasser, S. M. (1996). *J. Cell Biol.*, **134**, 1349.
24. Gotta, M., Laroche, T. and Gasser, S. M. (1999). In *Methods in Enzymology*, p. 663. 304: *Chromatin* (eds P. M. Wassarman, A. P. Wolffe), Academic Press, San Diego, CA.
25. Uzawa, S. and Yanagida, M. (1992). *J. Cell Sci.*, **101**, 267.
26. Straight, A. F., Marshall, W. F., Sedat, J. W. and Murray, A. W. (1997). *Science*, **277**, 574.
27. Guacci, V., Hogan, E. and Koshland, D. (1997). *Mol. Biol. Cell*, **8**, 957.
28. Jin, Q-W., Trelles-Stricken, E., Scherthan, H. and Loidl, J. (1998). *J. Cell Biol.*, **141**, 21.
29. Guacci, V., Hogan, E. and Koshland, D. (1994). *J. Cell Biol.*, **125**, 517.
30. Stone, E. M., Laroche, T., Heun, P., Pillus, L. and Gasser, S. M. (submitted).
31. Sherman, F. (1991). In *Methods in Enzymology*, Vol. 194, *Guide to yeast genetics and molecular biology*, (ed. C. Guthrie), p. 3. Academic Press, San Diego, CA.

4

Instruments for fluorescence imaging

W. B. AMOS

1. Introduction

Fluorescence microscopy is a highly sensitive method: in the last decade it has been extended to the detection of single fluorophores (1). To do this, as much as possible of the emission must be collected and a low background must be achieved by shielding the detector from the short-wavelength light used to excite the fluorescence.

There are many types of instrument for detecting fluorescence and, unfortunately, no single figure of merit for comparing them. Most researchers make their own independent comparisons, but a microscope, with its many controls and variable specimen, is the very worst test-bench. A plurality of instruments is currently in use, each with its own special advantages and none perfect. The purpose of this chapter is to compare the different types of instrument and to suggest criteria which might be used to arrive at a rational choice.

Two principal types of microscope are used for fluorescence imaging: wide-field (or conventional) and scanning. The wide-field category includes conventional epifluorescence and total internal reflection fluorescence (TIRF). Scanning microscopes include confocal, multiphoton and near-field (NSOM). The commonest types of confocal and multiphoton microscopes work by scanning a single spot of light over the specimen but multispot or slit-scanning forms exist, which are sometimes referred to as parallel microscopes since they consist of many confocal or multiphoton optical paths being used simultaneously and in parallel. The advantages of single-spot scanning for simultaneous imaging in fluorescence and differential interference contrast (DIC) are discussed below.

Figure 1 shows three forms of epifluorescence microscope. *Figure 1a* shows a conventional wide-field system in which light (usually from an arc lamp) is reflected into an objective lens and then illuminates a substantial area of the specimen. Some of the light emitted by the fluorescent specimen passes through a chromatic reflector towards the detector (a camera) which is shielded from any reflected exciting radiation by a barrier filter (not shown in the diagram). *Figure 1b* shows a scanning laser microscope. In this, a single spot of light is scanned over the specimen, or, more rarely, the specimen itself is scanned relative to a stationary spot. The purpose of this is to reduce the degradation of the image by light from regions of the specimen above or below the plane of focus. This is achieved by the use of confocal image-plane apertures or by multiphoton excitation (see below). Light is brought to a focus, normally as small as possible (to the limit set by diffraction over the aperture of the objective lens). Light emitted from the intensely illuminated volume at the focus is passed to a unitary detector (photomultiplier or avalanche photodiode). The detector response during scanning is used to build an image progressively in computer memory. The most widely used systems take approximately one second to build a complete image at acceptably high resolution (e.g. 512 lines). Paradoxically, although these systems are slow in the sense of framing rate, their integration time is short in relation to that of a television camera (a few microseconds as against 30 msec). This has led to their extensive use in physiological experiments, often in a line-scanning mode which offers time resolution of 2 msec.

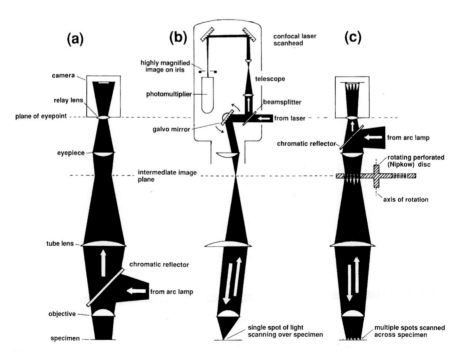

Figure 1. Three forms of epifluorescence microscope: (a) conventional wide-field, (b) confocal single-spot scanning and (c) parallel multiple-spot confocal system based on a Nipkow disc.

Many 'parallel' systems (*Figure 1c*) have been proposed or manufactured (see below) as attempts to improve on the single-spot type of scanner, particularly in framing rate. With these microscopes, instead of a single spot being illuminated, there are multiple spots or a thin line of light is used. Cameras are the only possible type of detector.

The single-fluorophore work of Betzig and Chichester (1) was done with a near-field scanning optical microscope (NSOM). This is completely different from the previous microscopes in that the single illumination spot is defined by the leakage of light through a hole of sub-wavelength dimensions in the tip of a light guide. The spherical wave emerging from the hole decays rapidly, which results in suppression of the exciting radiation and greater resolution than can be achieved by focusing light with lenses. This approach has been extended to the observation of fluorescence resonant energy transfer between individual pairs of molecules by Ha *et al.* (5) using a stationary probe which is imaged confocally on an avalanche photodiode (see below) while the specimen is scanned relative to the probe by a piezo mechanism. For NSOM to work, the specimen must consist of individual macromolecules or membranes, attached to a substrate such as mica or glass which is highly flat. NSOM will not be discussed further here, because it is not yet commercially available.

An important alternative to epifluorescence, which achieves a lower background, is total internal reflection fluorescence (TIRF), shown in *Figure 2* (2–4). Intense light at the excitation wavelength is totally internally reflected

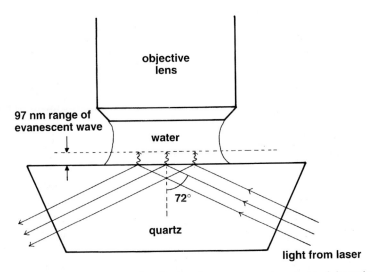

Figure 2. Essentials of total internal reflection fluorescence microscopy. A laser beam is directed to a quartz/saline boundary, to which the fluorescent specimen is very closely attached. The angle of incidence of the beam is kept at >65° (the critical angle) so that the light is totally internally reflected. The evanescent wave penetrates through the interface and can excite fluorophores within about 100 nm of the surface.

by a glass–water boundary. Fluorescence is excited by the non-propagated disturbance termed the 'evanescent wave' which passes through the boundary into the aqueous medium. The background is much lower than with conventional epifluorescence, particularly if silica (which has almost no fluorescence) is used instead of glass, but the fluorophore must be situated within 90 nm of the boundary; a serious restriction with cellular specimens.

Almost all these fluorescence detection systems rely on filters. Generally, it is necessary for these filters to have sharp cut-offs, which are achieved by multilayer dielectric construction and very strong blocking effects (optical densities > 5). For the latter, tinted glass remains the best method of construction and many filters are a sandwich of dielectric and bulk-coloured layers. A hand spectroscope used with a strong tungsten lamp illuminator is the best apparatus for checking such filters (for example, to check for leakage of red light through a green excitor filter designed for rhodamine): ordinary chemical spectrophotometers are not reliable for optical densities of >2. The catalogues of Chroma and Omega should be consulted for information about modern filter technology.

2. Light budget

A typical fluorophore used in biological studies can be excited repeatedly at rates up to 10^5 cycles/sec, but is destroyed , with a half-life of about 0.1–1 sec at such a rate. The chief problem of instrument design is to detect small bursts of photons. 10^5 photons/sec corresponds to the amount of light passing into the pupil of a human eye from a dark overcast night sky in which no stars can be seen. A useful comparison of scales of light units from which this fact is derived is given by Inoué and Spring (6).

Berland et al. (7) have discussed the light budget of a wide-field epifluorescence microscope. For best collecting power, the objective should have the highest possible numerical aperture (NA; see also Chapter 6), with the caveat that a nominal NA of >1.3 is pointless if the specimen is in water. Although the brightness of fluorescence objectives does increase with NA, another important determinant in epifluorescence is the area of the exit pupil: this tends to be as large as 8 mm diameter in objectives of around 60× magnification and results in the delivery of more excitation light to the specimen than is the case with a 100× objective of the same NA. Phase contrast objectives should be avoided for low-light-level fluorescence because they may be only half as bright as their bright-field counterparts. The post-objective optics should produce the minimum magnification on the camera compatible with the required spatial resolution (see below).

According to the figures of Berland et al. (7), the total collection efficiency of an objective is unlikely to exceed 27% (including losses due to low transmission as well as the limited solid-angle of capture). The chromatic reflector normally passes only 80% and Berland et al. give 50% for the transmission of the barrier filter and relay lenses. The camera will therefore receive only 11% of the photons emitted by each fluorophore before it is destroyed: this corresponds to 10^4 photons. Standard video-rate cameras cannot register this level of light, but scientific charge-coupled device (CCD) and other cameras can.

As light detectors, the three systems shown in *Figure 1* are very different. In the wide-field case (*Figure 1a*) the exposure time of the detector is long (~20 msec with a video camera, but often extended to tens of seconds in scientific integrating cameras). This makes this type of system suitable for detecting non-fluorescent signals such as bioluminescence. In contrast, the scanning systems have a dwell time of the order of 2 μsec, with a framing rate of only once per second. They are comparable to a flash camera with an exposure time of only a few millionths of a second. Such a system is totally unsuitable for bioluminescence work. For fluorescence excitation, the light is delivered to the specimen in pulses, with peak intensity reaching 0.5 MW/cm^2. At such intensities, photochemical saturation of the fluorophore may occur. However, bleaching is no greater than in wide-field, because the total average power delivered to the specimen is usually <1 mW. Apart from the advantages of confocal and multiphoton operation, and the high time-resolution mentioned above, there are several other benefits in scanning. A confocal system can be operated in full room lighting, to which it is quite insensitive. The raster-scanning allows zooming, panning and other operating modes which make the system optically more versatile. Also, it is possible that the pulsed radiation is less harmful than continuous radiation, perhaps because it allows the diffusion of free radicals away from the focused spot.

3. Comparing instruments

3.1 Quantum efficiency

To measure the sensitivity of an instrument, it would seem necessary merely to measure the smallest quantity of light energy that it can respond to. But light is quite strange (8): it arrives as photons of constant energy (at a given wavelength) and all the commonly used detectors, including the human eye, are capable of responding to individual photons or very small numbers of them. The difference between detectors is chiefly a statistical one, best described by the quantum efficiency (QE). In the case of television cameras and photomultipliers, a photon that falls on the detector may (or may not) generate an electron and the QE is the number of detectable electrons produced as a fraction of the number of incident photons. The QE is a number between 0 and 1. The principal detector materials are silicon (used in CCD cameras and photodiodes) and phosphors (used in photomultipliers and image-intensifiers). *Figure 3* shows the QE of several of these types of detector at different wavelengths. Unfortunately, the QE is not determined simply by the absorption spectrum of the material: it depends also on the geometry. For example, a thin silicon layer allows more infra-red photons to pass through without being absorbed, so the QE curve is truncated at the high-wavelength end relative to a thicker layer of the same material. In photographic film, the photon generates a silver atom from an ion, sometimes with the intermediate stage of absorption by a dye molecule. The QE for all such material (defined in a different way) is apparently less than 0.01 even at the wavelength of peak sensitivity (9). Even at high levels of illumination and in colour, film seems to have been rendered obsolete by the development of electronic cameras with more than 2000×2000 picture elements (pixels) (10).

Quantum efficiency can be expressed as radiant sensitivity in amps per watt (R_s) instead of electrons per photon. If the wavelength is λ nm,

$$QE = 1240R_s/\lambda$$

It would seem from *Figure 3* that the best detector for all purposes would be a silicon-based one, such as a CCD. However, the QE is not the only parameter

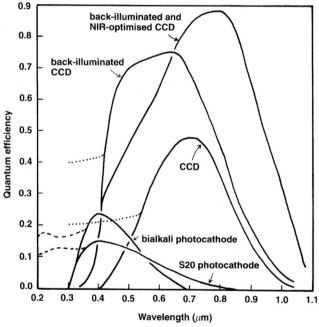

Figure 3. Variation of quantum efficiency with wavelength for a number of sensor materials. The dashed curves to the left of the photomultiplier curves for bialkali and S20 photocathodes show the ultraviolet improvement which can be obtained by the use of a quartz face-plate. The CCD curves are both from back-illuminated CCDs with different anti-reflection coatings. Extra ultraviolet sensitivity in CCDs (dotted lines) is sometimes achieved by coating with Lumogen, a fluorescent material (based on Holst (19) and also Thorn Vacuum Tubes data). The QE of the photomultipliers can be increased greatly by total internal reflection optics (see text).

that needs to be considered: when the light level is low, the noise in the detection process is equally important.

3.2 Noise

Noise is any feature of the output of a detector which does not carry information, but there are many different types of noise and their analysis has occupied many lifetimes of study.

At low light levels, photons fall on a detector in a random fashion, which can be described by Poisson statistics. The number of photons n, arriving during a definite period called the integration time, varies from one time interval to the next. The statistical variance is $\sqrt{n}$ and is called the 'shot noise'. Shot noise is a basic characteristic of light and nothing can be done to reduce it, though it can be made less significant by counting more photons (e.g. by increasing the integration time).

Electronic amplification also produces noise. A resistor produces thermal or Johnson noise, which is proportional to the absolute temperature. It can be shown that a capacitor introduces noise proportional to its capacitance. Most detectors produce a 'dark current' when no light is falling on them. Although it may be possible to subtract the average dark current from the current when light is admitted to obtain the correct signal, it is not possible to subtract the 'dark noise'. Noise from different sources is combined as total noise, N. For example, if the main contributors to the noise of a detector are shot noise (N_s) and amplifier noise (N_a) , then

$$N = \sqrt{(N_s^2 + N_a^2)}$$

Amplifier noise often dominates the performance of a detector at low light levels, but becomes irrelevant at high levels because the shot noise is so much higher. For high sensitivity, the amplifier noise must be minimized.

An important figure of merit is the signal-to-noise-ratio (SNR). If the signal from a detector is S and the total noise is N,

$$SNR = S/N$$

The SNR can also be expressed as a power ratio in decibels (dB):

$$(S/N)_{dB} = 20 \log (S/N)$$

(The reason for the 20 rather than 10 is that if S is expressed as a photo-current, the power is proportional to S^2.)

The SNR is often expressed in bits. The number of bits is $\log_{(base\ 2)} (S/N)$ or $(S/N)_{dB}$ divided by 6. Note that even when the SNR is expressed in bits, it is still a ratio of analogue quantities.

3.3 Dynamic range

A detector which could respond to single photons, but became saturated at very low light levels, would be difficult to use, except as a photon-counting device (see below). The range over which a proportional response can be obtained is called the dynamic range. It is defined as the signal level at saturation (SAT) divided by the dark noise of the detector.

$$\text{Dynamic range} = \text{SAT}/N_{\text{dark}}$$

This is roughly equivalent to the maximum signal divided by the minimum detectable increase. Under bright conditions, the human eye can detect an increase in brightness of around 2%, so, by the second definition, its dynamic range is only 50, or 6 bits. Standard video cameras have a range of 256 or 8 bits per frame.

If the integration time is increased, e.g. by averaging over 256 frames, the noise is reduced and so the dynamic range is increased by a factor of 16 (the square root of the number of frames). The total dynamic range then becomes $256 \times 16 = 4096$, or 12 bits. In all systems where the signal is digitised, the analogue dynamic range of the detector should be at least equalled by the range in bits of the analogue-to-digital converter. However, it is important not to confuse the two ranges: a 16 bit analogue-to-digital converter with a standard video camera will not generate 16 bit images, unless there is provision for averaging thousands of frames.

Because the eye is only a 6 bit device, it is impossible to make all the information in a 12 or 16 bit image perceptible at one go. However, it is often desirable to record images with high dynamic range. The range can be displayed in 6 bit pieces, and unsuspected detail may then emerge from the highlights or shadows. A high range is essential if extreme gamma transformations need to be made . Gamma is the function relating digital intensity to display brightness: it is often desirable to make this non-linear, for example, in order to devote proportionally more of the grey levels available in the display to the low-signal regions of an image. It is also essential to have a high dynamic range if brightness ratio calculations are to be made (as with ratiometric physiological fluorescent probes).

3.4 Photon counting

Provided the intensity of the light remains low throughout the period of observation, unitary detectors such as photomultipliers and some cameras can be operated in a photon counting mode. In this, advantage is taken of the quantal nature of light. The signal from the detector, viewed as a time trace, contains small peaks which are disregarded, as due to thermal effects, and larger peaks which are taken to denote the arrival of single photons. The signal is passed through a discriminator, which eliminates all peaks of below a certain height. The peaks are then turned into square pulses of absolutely constant height and length and passed into an electronic counting circuit. If the number of pulses per integration time is plotted on a histogram, the result (assuming that shot noise is the only cause of variation) is a Poisson distribution, in which it may be possible, by adjustment of the intensity, to identify the first peak as due to zero photons, the next to one per integration period and so forth. This way of handling the signal has the advantage that the background level can be identified unequivocally, the absolute level of the signal is easier to calibrate and non-linear detector responses and small errors in analogue to digital conversion have virtually no effect.

Using this mode, the absolute minimum level of illumination can be identified and used, so the photodamage to the specimen can be minimized. The only disadvantage of photon-counting operation is that the illumination level has to be set carefully, so that photons do not arrive too fast for the counter, or saturate the detector.

4. Getting the right magnification

In any instrument where an optical image is cast on to a detector, the magnification needs to be adjusted carefully. In a microscope, once an objective lens of the highest possible NA has been chosen, the magnification at the detector can be controlled by the use of relay lenses, such as that shown in *Figure 1a*. The intensity at the detector is proportional to the inverse square of the magnification, so low magnification is desirable, but not if it results in a loss of detail in the image. The simple formulae for calculating the optimum magnification are given next.

4.1 Magnification for cameras

The image of a point source of fluorescence formed by an ideal objective lens (together with the tube lens, if the microscope has one) is called an Airy pattern. It is formed in the intermediate image plane, shown in *Figure 1* panels a–c. It is a disc of light surrounded by concentric bright rings. For an objective of magnification m and numerical aperture NA, used with light of a single wavelength, λ μm, the diameter of the Airy disc (the distance between the first minima in micrometres) in this plane is given by

$$\text{diameter} = 1.22 \, \lambda m / \text{NA}$$

Assuming the use of green light with $\lambda = 0.5$ μm, the diameter is 13.6 μm for a 10× objective of NA 0.45 and 43.6 μm for a 100x, NA 1.4.

Figure 4 shows the calculated Airy disc sizes for these and other objective lenses superimposed on the pixel lattice (i.e. the array of rectangular photo-sensitive elements) of a camera. The camera format chosen has among the largest pixels currently available (20×16.7 μm). In order to capture all the detail from the image, the so-called Nyquist criterion has to be satisfied: the Airy disc must cover at least two pixels, preferably two pixels that are diagonally adjacent. Only the 100× passes this test: for the other objectives, the magnification must be increased. The extra magnification is usually achieved by means of a relay lens as shown in *Figure 1a*. If too high a magnification is used, the sensitivity will be reduced, although it can be restored in some cameras by combining the signal in adjacent pixels (pixel binning). If the pixel height and width are H and V in μm,

$$\text{extra magnification} = 2 \times \sqrt{(H^2 + V^2)}/\text{Airy disc diameter}$$

The pixel size for a range of camera formats is shown in *Table 1*. Sometimes a relay lens is required to give negative magnification, particularly with very small camera chips. In practice, many microscopists prefer to magnify more than is required to satisfy the Nyquist criterion, in order to record subtle changes in the intensity profile in the image. For example, in work with low-light-level fluorescence from live yeast cells, Silver (11) recommends a magnification which produces *c.*4 pixels per Airy disc. It should not be assumed that making the Airy disc larger than required by the Nyquist criterion is futile: if the SNR of the image is high, the position of an object in the microscope field can be determined with sub-nanometre accuracy by measuring the position of the centroid of the Airy disc (see, e.g., ref. 12).

4.2 Magnification in laser-scanning microscopes

In a simple flying-spot type of microscope, the spot is an Airy disc generated in the specimen by the objective, subject to lens aberrations. (To reduce possible saturation effects on the fluorescent dye, it would be desirable to increase the size of the focal spot in the special case where a display or computer memory of lower spatial resolution than the optical resolution is being used. This is seldom the case, however, and no commercial laser-scanning microscopes have this facility.)

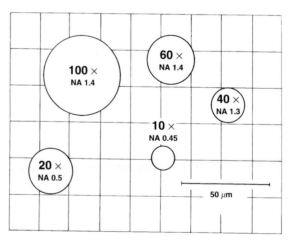

Figure 4. Airy disc sizes in the intermediate image plane for a number of commonly used objective lenses, shown superimposed on a CCD array.

Table I. Dimensions of arrays and pixels for CCD cameras (data from ref. 19)

Camera format[a]	Array size (mm)	Array diagonal (mm)	Nominal pixel size (μm)[b]
1 inch	12.8 × 9.6	16	16.7 × 20
2/3 inch	8.8 × 6.6	11	11.4 × 13.8
1/2 inch	6.4 × 4.8	8	8.33 × 10
1/3 inch	4.8 × 3.6	6	6.25 × 7.5
1/4 inch	3.2 × 2.4	4	4.17 × 5

[a]The camera format refers to the tube diameter of the now largely defunct tube cameras.
[b]The nominal pixel sizes are given for a 768 × 480 array.

In a confocal system, the illuminated spot emits light, by virtue of either reflection or fluorescence, and this light is is focused on to a circular aperture which ensures that only the light from the focused spot enters the detector. Provided the lenses are aberration-free, the emitted light forms an Airy pattern in the aperture plane. According to van der Voort and Brakenhoff (13), the best size for the confocal aperture in practice is approximately equal to the Airy disc diameter. Plainly, a different aperture size is needed for different wavelengths and different objective lenses. A key step in triggering the modern era of confocal microscopy was the demonstration by White in 1985 (see ref. 14) that if the magnification at the level of the aperture was increased greatly, a continuously adjustable photographic iris could be used with the same confocal performance as the pinholes used previously. The minimum diameter of the iris in millimetres is simply the Airy disc diameter (as calculated above) multiplied by the additional magnification factor. This factor can be supplied by the confocal microscope manufacturer: it is 50× for a Bio-Rad MRC series microscope (assuming that there are no extra magnification sources other than the objective lens) and slightly higher, (60x) for the Bio-Rad Radiance series. Thus, with a 60× objective of NA 1.4 the iris diameter should be 1.4 mm. The range of adjustment in the Radiance models is 0.7–12 mm, allowing the instrument to be set for ideal confocal performance (small aperture) or to have increased sensitivity with dim specimens with lessened optical sectioning ability by the use of a large aperture.

The magnification referred to here is the optical magnification of the image at the level of the detector iris. The on-screen magnification is quite different in a point scanning system, being determined by the size of the scanned raster. In such a system the magnification of the image can be varied independently of the optical magnification.

5. Light sources and fluorochromes

Arc lamps are the principal illuminant for camera-based systems, although a tungsten filament lamp may be bright enough for infra-red emitting fluorochromes when a sensitive long-integration CCD camera is used. The reason that the 100 W mercury arc is preferred to a 50 W mercury arc or a 1 kW street lamp is that the intrinsic brightness of the hot plasma inside the bulb is highest in the 100 W arc. When illuminating a small area, such as field of a 100× objective, the plasma is imaged at approximately unit magnification and the intensity is determined by the intrinsic brightness rather than by the size of the arc.

It is important to understand that the historic choice of fluorescein (a bright fluorescer) and rhodamine (less bright) was to balance the strong emission of the mercury arc at 546 nm (green) and its comparative weakness in the blue region.

If the strong emission lines of the mercury arc, or the tendency of the arc to flicker and wander, render it unsuitable, a xenon arc should be tried. This has a higher continuum radiation and is more stable. The spectra of arc lamps are given in the catalogues of suppliers (e.g. Oriel).

Laser scanning confocal microscopy is restricted in the wavelength of the source to the lines from available continuous wave (CW) lasers (see *Figure 5*). Unfortunately, the huge variety of wavelengths available from dye lasers cannot be used, because their pulsed output is unsuitable for scanned imaging. For a detailed discussion of multiple staining and laser excitation, see Brelje *et al.* (15). The web address of a useful database is http://www.fluorescence.bio-rad.com

6. Light detectors

The several types of light detector have different characteristics. No detector is suitable for all applications.

6.1 Photomultipliers

Although quite old, the photomultiplier (PMT) still dominates microscopy as the detector of low light levels with a fast response time. This makes it the detector of choice for scanning microscopy. The end-window type of PMT consists (*Figure 6*) of an evacuated enclosure, coated internally at one end with a thin, semi-transparent layer of phosphor, the photocathode. Photons absorbed by the photocathode cause electrons to be ejected which are attracted by a voltage gradient to a series of electrodes called dynodes, held at progressively higher voltages. The dynode series acts as an electron multiplier, since a single electron can induce the release of multiple electrons from each dynode. This process introduces multiplicative noise, but is superior in signal-to-noise level to most electronic amplifiers. The signal is taken off the tube as a photocurrent flowing into the anode, which precedes the last dynode. A gain of as much as 10^7 is possible with a PMT. It is possible either to operate this detector as an analogue device or to count photons (as described above). It is

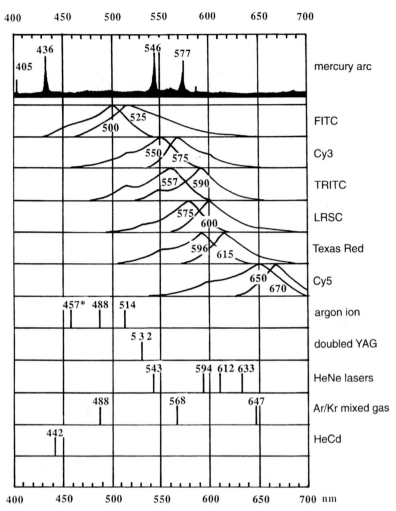

Figure 5. Spectra of a mercury arc, together with the absorption spectra of the principal dyes and lasers. *, Line available from lasers of 100 mW and above. LRSC, lissamine rhodamine thiocyanate.

not necessary to cool a PMT to use it in photon-counting mode although the dark current and dark noise are increased by higher temperature.

The disadvantages of PMTs are several. They show considerable batch variation in dark current and gain, and they can be damaged by high illumination, sometimes taking many hours to recover, if they ever do. The phosphors give them good visible and ultraviolet performance (forms with a quartz faceplate are available), but they are insensitive to infra-red. They need a stabilized high voltage supply. Even at their peak wavelength, two-thirds of the light goes right through the photocathode without stimulating the release of

(a)

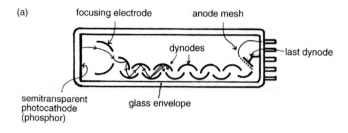

(b)

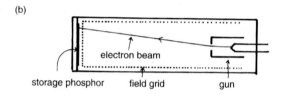

(c)

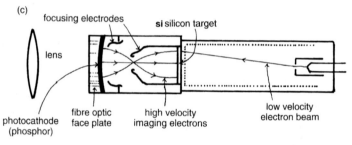

Figure 6. (a) Photomultiplier tube, (b) Vidicon and (c) silicon intensified target cameras. The focusing and scanning coils which surround the camera tubes are not shown.

electrons. It was discovered in the 1960s that this situation could be improved by subjecting the incident light to multiple reflections within the photo-cathode plus envelope (16) and a proprietary prism for achieving this, designed by the author, is used in the Bio-Rad confocal systems. This gives a useful increase in QE of approximately twofold in the green, fourfold in the visible red and higher in infra-red.

6.2 Photodiodes

Although photodiodes are small, cheap and electronically simple, their dark noise levels after amplification are too high for fluorescence imaging. Diode arrays have, however, been used in fluorescence macroscopy (see ref. 17).

6.3 Avalanche photodiodes

The avalanche photodiode (APD) is a relatively new type of detector. It is a silicon diode constructed and run in such a way that a single photoelectron can generate a cascade of electrons in a way reminiscent of a PMT, but the multiplicative noise is higher than in a PMT and the maximum gain when run as an analogue device is much lower (<1000 rather than 10^6). The dark current is probably a little lower than in a PMT, but the photosensitive area is much lower: a maximum of about 0.8 mm in diameter rather than 25 mm or more for PMTs. An APD can be run in both analogue and photon-counting modes. In the latter mode, the dark count and multiplicative noise have little effect. Attractions of the APD include its small size, high quantum efficiency (>0.8 at peak wavelength) and performance superior to a PMT in the red. Its spectral range reflects the silicon absorption spectrum, like that of CCDs (*Figure 3*). It has been used most effectively as a confocal detector by Weiss and colleagues (5).

6.4 Television cameras

Attaching a camera to a microscope is usually done by means of a threaded seating called a C-mount. A C-mount (on the microscope) has a male 32 t.p.i. thread with an external diameter of 1 inch. It screws into a female thread on the camera and the flat face on the male portion comes into contact with a corresponding flat on the camera. At this point the camera target should be 0.69 inches from the flat face but often it is not. Some cameras have adjustments for this (see Inoue and Spring (6) for a drawing of a C-mount with dimensions). A trend towards smaller camera chips and lenses of shorter focal length has resulted in so-called CS cameras. These should not be attached to a microscope C-mount without a special spacer, which increases the C-mount-to-chip distance by 5 mm. Three-chip cameras (see below) often have non-standard mounts.

When attaching a camera as in *Figure 1a*, it may be necessary for a user to find a suitable relay lens to produce a required degree of magnification. A simple achromatic doublet lens is sufficient, and should be placed in the plane of the eyepoint (the aperture plane just above the eyepiece where the light, in a bright-field microscope, is focused to a small disc). Camera lenses are usually not suitable for this purpose, because the entrance pupil of the lens (where the lens iris is situated) is too deep inside the lens for it to be made to coincide with the plane of the eyepoint of the microscope. Zoom lenses are not advised for fluorescence work, because their light transmission is usually poor. If the magnification required is large (as in video-enhanced DIC) no relay lens is necessary: the camera target is simply placed at a distance of ≥150 mm from the eyepiece and the microscope focus can be altered slightly to bring the image into focus. This use of the objective slightly further from the specimen than normal would be expected to introduce spherical aberration but actually works quite well.

6.4.1 Vidicon tube

The type of television camera used throughout the growth of television was the Vidicon tube (*Figure 6*). In this, an image is thrown by a lens on to a storage phosphor, where charge is locally accumulated at points of higher illumination. The charge is released by a scanning focused electron beam similar to that in a cathode ray tube or television monitor which is controlled by magnetic coils (not shown in the diagram). As a result, the Vidicon is able to integrate the light input over the period (about 20 msec) between scans and produces a time-variant electronic signal organized into line and frame scanning segments (see refs 6 and 8 for detailed and comprehensive accounts of video technology). Although tube cameras such as the Newvicon are still extensively used in microscopy they are not sensitive enough for fluorescence work.

6.4.2 SIT cameras

An image intensifier is an evacuated enclosure with a phosphor coating at one end. As in a PMT, electrons are ejected from the phosphor and accelerated by a high voltage gradient, but, unlike the PMT, the image information is pre-served: the electron trajectories are carefully shaped so that an electron image similar to the original light image is formed on a second layer of phosphor. The second phosphor emits light, showing an image which can be much brighter than the original one. Devices like this were used for gunsights and other military purposes. A silicon intensified target (SIT) camera (*Figure 6*) has an image intensifier in front of a tube camera, but instead of generating light by means of a phosphor, the intensifier delivers the electron image to a thin plate of silicon, the target. The scanning beam scans this silicon target instead of the storage phosphor. SIT cameras have considerably higher sensitivity than simple tube cameras. They are used for fluorescence imaging and are cheap but are not ideal. Their faults include a lag induced by the phosphor, which results in some unavoidable image integration (which may make the image look good), a poor dynamic range and tendency to saturate. They also have the besetting fault of all tube cameras: a tendency for the scanning raster to vary with time and to have imperfect shape, so that geometrical distortions occur in the image. An ISIT is a SIT camera with another intensifier in front of the basic one.

6.5 CCD cameras

CCDs are the most commonly used television cameras at present (see Chapter 5 for a discussion of CCD camera selection for imaging living cells. They combine technology developed for digital registers and phototransistors in a form which is compact, cheap and totally free of geometrical image distortions.

The way in which a CCD works is shown in *Figure 7*. A rectangular array of transparent rectangular electrodes is formed by photofabrication on a thin wafer of silicon coated with silicon oxide. Electrical connections are made to each electrode, as shown by the leads 1, 2 and 3 in the diagram. In the terminology of transistors, each electrode is called a gate and controls a metal oxide/silicon (MOS) junction adjacent to it. When a photon hits the silicon, the resulting electron is trapped at the junction between the silicon oxide and *p*-type silicon. Electrons can accumulate in this way throughout the integration time of the device, which is controlled electronically and can be varied from microseconds to tens of milliseconds (video rate) or extended for tens of seconds, as is often desirable in fluorescence recording. Instead of being read out by an electron beam in a vacuum tube, an elegant charge-shifting procedure is used. Let us suppose that at time zero (*Figure 7*) , electrons have accumulated under gates G1 and G4, the other gates being protected from light by an opaque coating over the electrodes. A voltage is applied to line 2 so that the electrons spread (still trapped at the junctional interface) underneath G2 and G5. Then, by changing the voltage on line 1, the electrons can be shifted totally under G2 and G5, and the cycle can begin again. A similar process can shift the packages of electrons in a direction perpendicular to the paper to that the charge variation can be fed into an amplifier to produce the

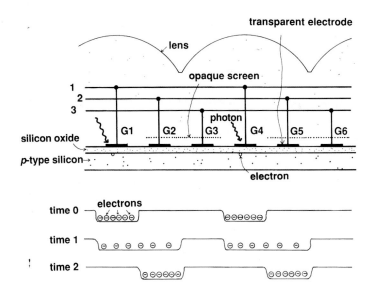

Figure 7. Diagrammatic section through a CCD array (see text). The time series in the lower half of the figure shows how electrons trapped at the oxide/silicon boundary are shifted by manipulating the voltage on lines 1, 2 and 3.

signal. Note that only the gates G1 and G4 must be illuminated, so only they correspond to pixels in the image. As a consequence, only a small fraction (called the 'fill factor') of the surface of the CCD is photosensitive. This factor can be improved by adding microlenses, as shown in the diagram, to channel more light on to the photosensitive gates.

CCD chips from a small number of suppliers are incorporated into cameras by a multitude of manufacturers, with the principal variation being in the electronics, e.g. in the method of integration of signal. Among monochrome chips (see *Table 1*) there is variation from tiny arrays designed for low-resolution video to very large and costly arrays of 2000×2000 pixels or more, which are made chiefly for astronomy and aerospace applications. For scientific purposes, 1000×1000 is now quite commonplace.

There are several types of CCD, differing principally in the way the charge is read out (see ref. 19 for a detailed account of CCD devices). The simplest type is the full-frame device in which the charge pattern is transported *en bloc* to the edge of the array, where the image is placed, a line at a time, in a serial readout register. A mechanical shutter is necessary to prevent light falling on the array while the pattern of charges is being transported across it, which would have the effect of smearing the image. The next step in complexity is the frame transfer device, in which a complete frame of data is transported in about 0.5 msec into an array situated alongside the photosensitive array but shielded from light. The chip can be operated without a mechanical shutter but some smearing may still occur, particularly if the light intensity is high. Many modern CCDs are of the interline transfer type, in which a line of shielded CCD elements is placed next to each line of photosensors. It takes only 1 µs to transport the charge into this shielded storage, so very fast framing is possible. Because so much of the area (perhaps as much as 80%) is now not sensitive to light, it is important to have microlenses to improve the optical fill factor. The acronym HAD signifies a type of chip in which holes rather than electrons function as the charge carriers. This apparently makes it possible to empty the storage elements totally between readouts and eliminates image lag.

Colour imaging can be achieved by having a single chip with the pixels covered by a mosaic of colour filters. To achieve greater sensitivity chiefly by improving the fill factor, three CCD chips can be included in one camera, with red, green and blue images formed on them by directing the light through a chromatic splitter consisting of a set of prisms. 'Three-chip cameras' are more costly than those with single colour chips, but are definitely superior for multicolour fluorescence applications.

6.5.1 Video-rate CCDs and readout noise

Ordinary video-rate CCD cameras are not, in general, suitable for low light imaging. This is because their dark noise and readout noise reduce the effective sensitivity too much. Readout noise arises because the charge shifting described above cannot be carried out quickly without some residual charge being left behind. Cheap cameras designed for surveillance are quite useful as laboratory viewers for work involving infra-red: many cameras have an infra-red filter built into them to reduce this sensitivity.

6.5.2 Long integration CCDs

If the integration time is increased to seconds or tens of seconds the readout noise becomes insignificant, because there is only one readout operation and it can be carried out in a slower and more orderly manner. The dark current can be reduced by lowering the temperature of the CCD: it is reduced by a factor of two for each 8–9 °C of cooling. The usual method for cooling the chip is to use the Peltier effect, in which heat energy is extracted by an electric current passing between dissimilar metals and released into a heat sink elsewhere in the circuit sometimes with cooling water flowing through it. In order to avoid condensation, the cooled CCD array may be placed in a vacuum chamber.

The cooled CCD is a remarkably perfect imaging device. Large arrays, in excess of 1000 × 1000 pixels, are available. These provide a resolution far in excess of any video device, but their images can be stored manipulated and viewed by means of a computer. The cooled CCD is suitable for the most weakly fluorescent specimens and it works over an enormous dynamic range (e.g. 1:10000, 80 dB or 13 bits).

The gate capacity or 'well size' for a typical chip is 150000 electrons; the dark noise when cooled is <5 electrons r.m.s., and the dark current may be so low that it does not saturate (i.e. fill the well) for >100 sec. Holst (19) shows how to calculate the degree of cooling required to increase the saturation time to a given value.

The disadvantages of the cooled CCD are that it is not much improved over a conventional CCD when operated at high framing rates such as video rate, because of the high readout noise. Many cameras will provide images with long integration times only, so they are difficult to focus. The problem is compounded if an infra-red emitting fluorochrome such as Cy5 is used, which is invisible to the eye. To overcome this problem, some new cameras have the option of operating at video rate when required, for setting up the specimen.

6.5.3 ICCDs

If it were not for the readout noise, CCDs would satisfy all the imaging requirements for cameras. However, they do not perform well when the integration time has to be short (frequent readouts) and the signal level is also low. This occurs, for example, in following individual fluorescent molecules in motion, such as actin filaments labelled with rhodamine, which move at a rate of micrometres per second so that near-video rate is essential to follow the movement. What is needed is an amplification stage to increase the light signal before it reaches the CCD. Such cameras are called intensified CCDs (ICCDs).

An ICCD is shown in *Figure 8*. The optical image is cast on a phosphor layer inside an evacuated enclosure, as in the image intensifier described previously. In most modern cameras the intensifier is of the microchannel plate type. This consists of a thin plate fabricated of millions of glass tubes 10 μm in diameter, all set at a small angle to the perpendicular to the plate surface. This remarkable structure is made by heating a stack of capillary tubes and drawing out the stack to reduce the diameters before slicing it at the required angle and coating the interior surface of the tubes with metal. A voltage of between 600 and 1000 V is applied across the tube and a photoelectron entering at one surface is accelerated and releases more electrons as it passes through the tube, much like the multiplier effect in a PMT. At the exit surface of the plate is a second phosphor, which produces a light image considerably brighter than the input image.

The remaining problem is to couple the intensifier optically to a CCD. To do this with lenses is possible—e.g. a pair of low *f*-number camera lenses placed back-to-back as a 1:1 relay (*f* number = $1/(2NA)$)—but even carefully designed relay lenses have poor efficiencies, often as low as 2%. A better solution

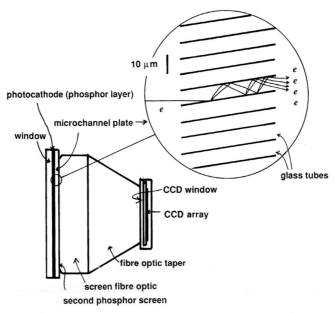

Figure 8. Intensified charge-coupled device (ICCD) camera. Light is first converted into electrons by the phosphor of the photocathode. Electrons generated by individual photons are multiplied in the microchannel plate (see text) and produce in turn a high output of light from a second phosphor. Unfortunately, this often needs to be reduced to fit a CCD chip: in this example by the use of a tapered coherent fibre optic block.

is to use a coherent fibre optic image-conduit. This is a block of optical fibre material which is placed close to both the phosphor and the CCD. Unfortunately, the fibres can accept light only over a limited solid angle (set by the fibre's NA) whereas the phosphor is a Lambertian emitter (i.e. it emits in all directions): as a result, the efficiency is only about 10%. If a tapered fibre optic is used to reduce the image to fit a small chip (as in *Figure 8*) there is a further loss, corresponding to the reduction factor of the cross-sectional area. (Avoid the common misconception that tapered optical fibres can concentrate light: even if losses into the cladding could be prevented by making the cladding totally reflective, a long tapered fibre would reflect light back toward the origin such that its throughput would be proportional to the reduction factor of its cross-sectional area. Welford and Winston (20) deal with the more complex question of light concentrators which are short in relation to their apertures.)

In spite of these difficulties the ICCD is a successful camera, much used for moving weakly-fluorescent specimens. It has the problems that the intensifier produces a high background 'snow' due to thermal effects on the phosphor and the intensifier is easily saturated. The dynamic range is poor and long integration times give low-contrast images with a big contribution from the background. Modern ICCDs have protective electronic devices to cut off the high voltage if the intensifier saturates, so damage to the highly expensive camera is avoided. Nevertheless, it may be extremely exasperating to search over a specimen with dark and bright regions, since the image 'whites out' frequently and the camera is forever cutting out. These problems have been reduced greatly in recent models.

6.5.4 CMOS cameras

In future, most mass-produced television cameras seem likely to be of a new type called the CMOS camera in which each pixel has its own on-chip amplifier. These cameras will be cheaper and more compact than CCDs, since pixel clocks, amplifiers and signal-processing devices are all fabricated on the chip. However, they are expected to have higher dark currents and lower fill factors than CCDs, so they will probably not render CCDs obsolete for fluorescence imaging.

6.6 Camera-based spectral imaging

The simple and familiar epifluorescence microscope shown in *Figure 1a* is sometimes equipped with multiple filters which can be exchanged mechanically (see Chapter 5). In addition, a number of devices have been invented which use a camera to collect spectral information from fluorescent specimens without using filters. None of these devices are widely used at present.

A 'pushbroom imager' (e.g. the Pariss system being developed by Lightform Inc. Belle Mead, NJ, USA), uses a slit in the plane of the intermediate image in conjunction with a prism which spectrally disperses the light at right angles to the direction of the slit. On the camera target lies a series of images of the slit each corresponding to a certain wavelength. A series of such camera fields is recorded, while scanning the specimen past the slit. In this way a complete spectrum can be recorded for every feature in the microscope image.

A somewhat more economical, though optically much more complex system, is the Fourier imaging spectrometer (21). This works by passing the light from the entire fluorescent field through a Sagnac interferometer and forming an image of the microscope field by uniting the two beams which have passed through the interferometer. The path-length difference in the interferometer is varied (by tilting a mirror) and a series of camera images of the field is obtained. The plot of intensity at a given pixel against path difference is called an interferogram: it would be a sine curve for a pixel receiving monochromatic light. Methods for obtaining the emission spectrum from the interferogram by reverse Fourier transformation are well known in spectrometry. This remarkable method comes into its own when applied to preparations with six or more different fluorescent stains. There is naturally great interest in this for medical karyotype analysis (see ref. 22 for an alternative, possibly more efficient, apparatus).

6.7 Testing cameras

Although few biologists have time to do it, cameras can be compared by imaging a standard resolution target (e.g. the USAF pattern) with a normal camera lens (preferably one designed for field sizes of a few centimetres). Apparatus for this purpose is available commercially, but can be improvised by equipping a light-proof box with a bulb, a diffuser, a target and a series of neutral density filters and possibly also chromatic filters. This is fine for long-integration time CCDs but does not give much guide to the performance of ICCDs with moving targets. Comparing cameras using biological fluorescent specimens in a microscope is fraught with difficulties because of specimen variability. Also, it must be borne in mind that cameras may be strongly affected by infra-red or low levels of ambient light that cannot be seen, including light entering through the binocular eyepieces of a trinocular assembly.

7. Scanning optical microscopes

The chief types of scanning optical microscopes for fluorescence applications are confocal microscopes of the point scanning and parallel types (illustrated in *Figure 1*) and multiphoton microscopes.

7.1 Advantages of scanning *per se*

Some advantages of using a scanning system have already been mentioned above, such as the higher time resolution than in a video camera system.

Another is the ease with which multiple imaging modes can be combined: as the scanning spot passes over a feature in the specimen it can send signals to any number of detectors, including epifluorescence, transmission and epi-reflection. Since the location of these signals in all the resulting images is determined temporally, perfect co-registration is automatic and is not dependent on precise positioning of the detector. To microscopists unfamiliar with this idea, it is instructive to hold the transmission detector head in a laser-scanning imaging system, to tilt it or move it off axis and see there is no change in the scanned transmitted light image, except in intensity.

Another advantage of laser-scanning microscopes is that the laser beam is normally linearly polarized. This makes it unnecessary to interpose an analyser between the scan head and the specimen, where it drastically reduces the epi-fluorescence signal. *Figures 9* and *10* show how a laser scanned transmission image is obtained with DIC without the use of an analyser.

It is much more difficult to obtain simultaneous low-light-level fluorescence and DIC images in a conventional camera-based system. Foskett (23; see also 24) devised one solution, which is to use a chromatic beam splitter both to separate the light for the DIC image from that for fluorescence and to serve as an analyser. This works well, but places limits on the wavelengths that can be used and, since two cameras must be used, has the problem that the two images must be brought into register carefully by mechanical and/or electronic means.

7.2 Confocal microscopes

A confocal microscope (*Figure 1b*) is one in which the illumination is focused on a region (spot or line) in the specimen and the detection is confined to the same region. Confocal images can be produced only by scanning the spot over the specimen or vice versa. There are two distinct advantages in confocal operation. With any size of focused spot and detector aperture, geometric restriction gives an optical sectioning effect, in which unfocused light from outside the plane of focus is reduced. But if the input and output apertures are both imaged as one diffraction-limited spot, the resolution (see below) is improved. Because of the clarity of the optical section images, use of the confocal laser-scanning microscope has expanded over the last 12 years to the point where it is now a standard laboratory tool.

7.2.1 Confocal microscopes where a single spot of light is scanned

The scanning stage design of the early 1950s is the oldest (see ref. 25). In this, the beam and optical elements are kept stationary. This is optically good in that only the centre of the field of the microscope objective is used, where aberrations are at their lowest. Although this has not been widely used because of problems with vibration and slow scan speed of the early models, it is undergoing a revival, notably for NSOM in the single-molecule work previously referred to (5).

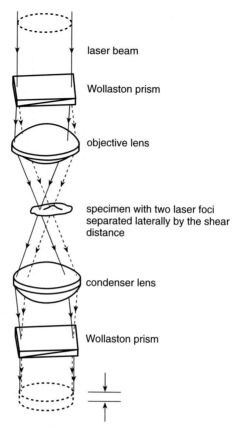

laser beam

Wollaston prism

objective lens

specimen with two laser foci
separated laterally by the shear
distance

condenser lens

Wollaston prism

phase difference (constant over entire
condenser aperture but varying with scan
position on specimen)

Figure 9. Scanning DIC using an upright microscope. The laser light passes through the optics in the reverse of the usual direction, and forms two focused spots in the specimen, displaced from each other along the direction of shear. The phase difference between these two produces an intensity which is constant over the condenser aperture, but varies with time. This variation is invisible unless a polarizer is inserted between the condenser and detector.

Most of the confocal microscopes in current use employ a single-spot scanning system using laser light as the source and mirrors placed in aperture planes to achieve the scanning motion. Lasers are available only at certain specific wavelengths some of which are shown in *Figure 5*. Arc lamps cannot be substituted because they are not bright enough after collimation (which is essential for producing a diffraction-limited spot) and are even dimmer after passage through an excitor filter.

The essential parts of a modern single-spot confocal scanning system are shown in *Figure 11*, which is based on the Bio-Rad Radiance series. Note that

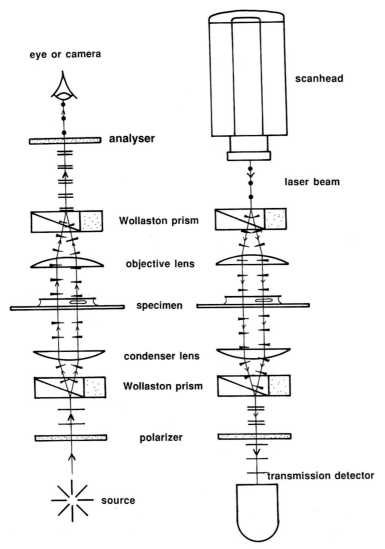

Figure 10. This illustrates the superiority of a scanning laser microscope over a camera for imaging simultaneously in DIC and fluorescence. A camera (left) requires both a polarizer and an analyser to be used for DIC imaging. Since it can be arranged that the laser beam from the scan head (right) is plane polarized in the correct azimuth, no analyser is required and fluorescent emission can pass at full strength into the scan head. The same laser wavelength as is used for exciting fluorescence can be used to form the scanned (but not confocal) DIC image. The diagram is intended to show that, with the initial laser polarization north–south (perpendicular to the paper), the Wollastons are oriented at 45° to the polarization direction and the polarizations of the rays in the central part, between the Wollastons, is at 45° to the plane of the paper and mutually perpendicular in the two ray-paths. After the polarizer (which is essential for both schemes) the laser light passes into a unitary detector, a photomultiplier or photodiode.

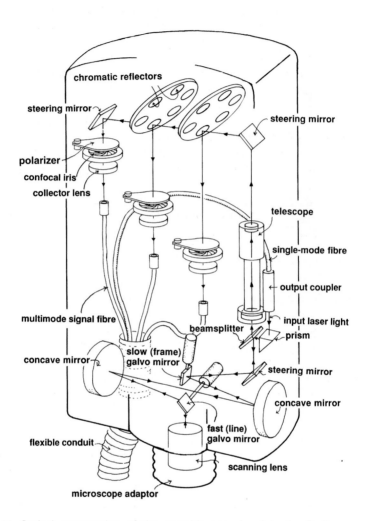

Figure 11. Optical construction of the portable scan head in the Radiance series of confocal microscopes (Bio-Rad). Laser light is introduced through a single-mode fibre and 30% reflected off a beam-splitting plate (more efficient in reflection because of the direction of polarization of the laser). A steering mirror guides the light into a patented mirror-based scanning system from which it is scanned in two perpendicular directions into a scanning lens similar to an eyepiece. Light from the illuminated spot in the specimen is descanned by the same system and 80%, if unpolarized, passes through the beam splitter and into the telescope. The scan lens and telescope produce an additional magnification of approximately 60× at the level of the confocal iris. The polarizer is switched in for non-specular reflectance imaging (e.g. immunogold) and, in addition, a quarter-wave plate is added to the microscope for specular reflectance imaging (e.g. interference–reflection). The collector lenses gather the light into the multimode signal fibres, according to wavelength and the signal is carried to a remote detector unit housing the barrier filters and photomultipliers with prismatic enhancers.

the scanning system consists of two small mirrors which are caused to oscillate by a mechanism resembling a moving-coil galvanometer (hence the term 'galvo mirrors'). These are optically imaged on each other, forming an all-reflective scanning system (26). Instruments using this system have been modified to scan equally rapidly in any direction, so the image can be rotated as desired. For example, a neuronal axon, whatever its real orientation, can be positioned parallel to the fast-scan direction, so giving the best time resolution for physiological experiments. The beam returning from the specimen is, like the input laser beam, nearly parallel. The function of the telescope is to focus the specimen, at an enlargement of approximately 60 times that of the inter-mediate image, in the plane of the detector iris, in such a way that the image of the illuminated spot is centred on the iris. In some earlier models the magnification was achieved by having an optical path of approximately 1.5 m folded inside the scan head. By adjusting the diameter of the iris, the best compromise between confocal optical sectioning and signal strength can be achieved, as described in section 4.

In this design, the emitted light is split into fractions of differing wavelength by means of the chromatic splitters and these fractions are directed ultimately into different detectors. There is a separate, servo-motor-controlled iris for each of these channels. It is desirable to have separate irises for each of the channels since the Airy disc diameter scales with wavelength (see above) and since one of the stains in a triple-stained preparation may be very dim in comparison with the others and may need a wider iris. A separation of the PMT detectors from the scan head is achieved by the use of multimode high-efficiency optical fibres with a large core diameter: this allows the scan head to be conveniently small and avoids heating of the PMTs by the galvo coils and head electronics.

In the above apparatus the maximum scan rate is set by the energy dissi-pation in the galvo mechanism at 750 Hz. Rapid framing (e.g. 16 frames/sec) can be achieved, but only by reducing the image resolution to 50 lines instead of the normal 512 or 1024. In a more complex apparatus (27) the fast mirror is run at its resonant frequency (adjusted to 7.8 kHz) and used bidirectionally so that video framing rate (30 frames/sec) is attained.

7.2.2 Acousto-optic scanning

Draaijer and Houpt (28) demonstrated that higher speeds can be achieved by the use of a solid-state acousto-optic deflector (AOD). These consist of a solid block of glass or crystalline material which is converted into a three-dimensional diffractor by setting up a standing acoustic wave in it. The acoustic wave is generated by a piezo pushing device bonded to the side of the diffracting medium and energized by a radio-frequency electronic input. By ingenious design, almost all of the input monochromatic laser light can be dif-fracted into one first-order beam, and the beam can be made to scan through an angle by altering the frequency of the electronic control signal. The chief

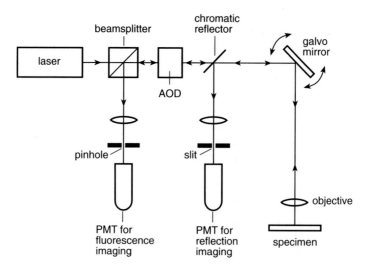

Figure 12. Confocal microscope of Draaier and Houpt, using an acousto-optic deflector (AOD) for the fast scanning axis and a galvo mirror for the slow. The fluorescent signal is not passed back through the AOD: instead it is passed to a slit aperture in a partially descanned state as a point focus which oscillates up and down the slit, but is detected at all positions by the photomultiplier (PMT). A second PMT is used to detect the reflection signal in a fully-descanned state with the use of a conventional confocal pinhole aperture (see text).

drawback of this kind of scanner is that it is, like any other diffracting device, dispersive: if the polychromatic emission from a fluorescent specimen were sent back through the AOD it would be spread and could not easily be gathered into a confocal aperture as it is in the mirror-based devices.

To solve this problem, Draaijer and Houpt designed the system shown in *Figure 12*, in which one axis (the frame scan) is achieved by a galvo mirror and the fast axis (line) is scanned by the AOD. The beam of fluorescent light emitted by the specimen is descanned in one axis by the mirror and is diverted by a chromatic reflector to a PMT. Since it is still scanning in the orthogonal direction, semi-confocal optics requires the aperture to be a slit: the spot of light oscillates up and down this slit. Surprisingly, the images are of good quality, without obvious anisotropy. A reflected-light image can be formed by passing the monochromatic laser light back through the AOD to a second detector, which has a circular confocal aperture of the usual type.

Contrary to early predictions, AOD microscopes have not made the mirror-based ones obsolete, possibly because of technical problems in maintaining stability at high scan speeds and high data rates.

7.2.3 Parallel confocal microscopes (multi-spot or slit scanners)

All these microscopes use a camera as a detector. In one type, first designed by Petran (29), these microscopes use multiple apertures in a spinning disc

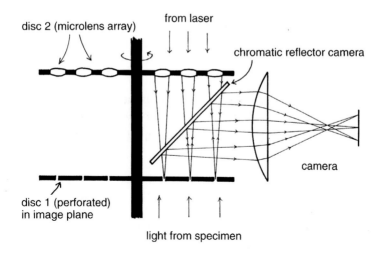

disc 2 (microlens array)

from laser

chromatic reflector camera

camera

disc 1 (perforated)
in image plane

light from specimen

Tanaami (Yokogawa Electric Co.)

Figure 13. Nipkow disc confocal multispot microscope of Tanaami and colleagues, showing the use of microlenses to improve the light throughput of the Nipkow disc.

(Nipkow disc) as shown in *Figure 1c*. The disc is placed in the intermediate image plane of the microscope and a large number of spots of light is focused on the specimen. The light emanating from each spot passes through the particular aperture from which it came (or, in Petran's original design, an equivalent one). When the disc is spun at high speed and its surface is viewed by means of an eyepiece, an apparently continuous image is obtained. The image is traversed by scan lines, but these are very inconspicuous in the best examples of this type of instrument.

Early forms of the Nipkow instrument were seldom used for low-light-level fluorescence imaging, because only a small proportion of the light of the arc lamp could pass through the disc. Recently, Tanaami *et al.* (30) improved on the Petran design by incorporating a second disc, equipped with a microlens array (*Figure 13*). Using a laser, the small holes in the Nipkow disc can be illuminated more efficiently because each microlens acts as a concentrator.

Another form of parallel confocal microscope uses a slit as the aperture, both for detection and illumination. As with the AOD scanner described above, the image, surprisingly, shows little anisotropy. A simple and potentially inexpensive design with an array of slits, analogous to the simplest type of Nipkow system, was described by Lichtman *et al.* (31). Efforts have also been made to use separate and stationary slits for the illumination and detector apertures, to extend to parallel scanning the advantage of the variable aperture which has been so clear with single-spot scanners. Such instruments have been designed by Brakenhoff and by White and Amos and manufactured commercially (see ref. 32). Although they have the advantage of rapid scanning at video rate, useful for visual searching and following moving objects, they have not proved popular for standard low-fluorescence biological specimens. There are three reasons for this:

(a) the slit scanning instruments are inferior in their optical sectioning properties, especially when the fluorescence consists of large continuous volumes;

(b) they are almost as expensive as spot-scanning instruments;

(c) they bleach the specimen quickly.

The limitations of epifluorescence confocal microscopy are shown in *Figure 14*. A serious practical limit is set to the useful scanning rate by the rate at which the fluorochrome can return to its ground state. If the intensity is high, the number of fluorochrome molecules in the ground state falls, with a resulting fall in optical density. In the extreme where the proportion of molecules in the ground state is negligible, the fluorescence process is said to be saturated: no extra emission occurs if the intensity is increased further. However, it has proved possible in practice to avoid saturation by controlling the laser intensity carefully. A more serious problem is bleaching, since when the fluorochrome

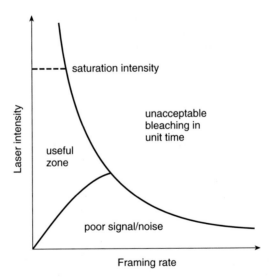

Figure 14. The dilemma of confocal microscopy. Absolute units are not given, because they vary according to the nature of the specimen. The 'intensity' referred to is that of the laser before passage through the microscope optics. Note the 'fatal hyperbola', to the right of which the bleach rate is unacceptable; it is hyperbolic because the bleach rate is the product of the frame rate and the laser intensity. A multispot scanner can be used to give higher framing rates, dividing the laser light so that each individual spot has a lower intensity, but it does not change the position of the hyperbolic boundary.

is bleached phototoxic products such as singlet oxygen are released. Bleaching is related to the total dose of radiation and is therefore proportional to the product of the laser intensity and the framing rate. Parallel confocal scanners can use lower laser intensities for a given data rate, but their *raison d'être* is faster scanning, where a higher rate of bleaching cannot be avoided.

7.2.4 How to test a confocal microscope

The following test specimens have been found to be most useful (see also Chapter 6).

(a) For testing axial resolution, a slide with a coverslip of measured standard thickness (0.17 mm) and aluminium evaporated on to the underside of the coverslip is imaged in reflection. Apparatus of the type used for carbon-coating in electron microscopy can be adapted to deposit aluminium, using a tungsten filament around which has been wrapped a small quantity of aluminium foil. To aid focus, many small scratches can be made in the aluminium by jabbing a paintbrush repeatedly at the film. A series of images (a z-series) is taken at different positions of the microscope focus and the intensity is measured in a small region as it appears in each image. The plot of intensity against focus position gives a peak, of which the width at half-maximum height is a measure of the axial resolution. Care must be taken to measure from a true baseline: use of an electronic offset can give false measurements.

(b) For measuring the point-spread function of the objective (i.e. the apparent distribution of intensity recorded by the instrument when presented with a point source), use sub-resolution fluorescent beads (e.g. Polysciences Fluoresbrite 210 nm or lower). The tricks are to view them close to the coverslip and suppress Brownian motion. A good method of preparation is to dry the beads down on the coverslip, which should cause the majority to adhere, and then invert the coverlip over a drop of polyvinyl alcohol (PVA)-based mountant (e.g. Gelvatol; see Chapter 1, *Protocol 9*), as described in Chapter 6, *Protocol 1*. Further details about point-spread functions are given in Chapter 6.

(c) For testing chromatic aberration, it is best to use beads which fluoresce in a broad spectral range and try to merge the images from spectrally different channels. Failure to merge with perfect superimposition indicates chromatic aberration, usually in the objective. Lateral aberrations are usually low, though they tend to increase in effect with distance from the centre of the field. Longitudinal aberration is present to some extent in all objective lenses. It is wise to do this test before embarking on a double-labelling immunofluorescence study.

7.2.5 Choosing an objective lens

As *Figure 15* shows, the depth resolution of objectives when used confocally improves approximately as the square of the numerical aperture of the lens.

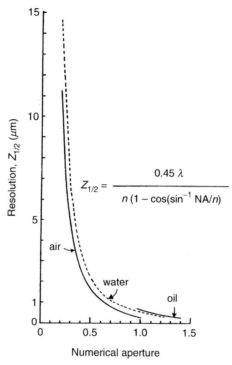

Figure 15. Axial resolution of a plane reflector by the FWHM criterion (see *Box 2*), according to a theoretical equation for confocal microscopy provided by Xiao and Kino (48). Note that for a given NA, a dry lens gives better axial resolution than oil or water immersion.

The best lens is one with the highest NA and the longest working distance (see *Figure 16*). If you are looking at living or aqueous specimens >40 μm thick, you will benefit from the use of one of the new (and highly expensive) water immersion lenses of NA 1.2: unlike oil-immersion lenses these show little deterioration of resolution on focusing deep into an aqueous specimen beneath a coverslip. An alternative approach, using immersion oils of differing refractive indices, is described in Chapter 6.

7.2.6 Does a confocal microscope have better resolution than a conventional one?

Confocal optics can effect an improvement of 40%. However, this improvement is seen only under ideal conditions: the big advantage of confocal imaging for epifluorescence is the elimination of out-of-focus glare, not better resolution. Moreover, the resolution criterion (*Box 1*) must be specified carefully (see *Figure 17*).

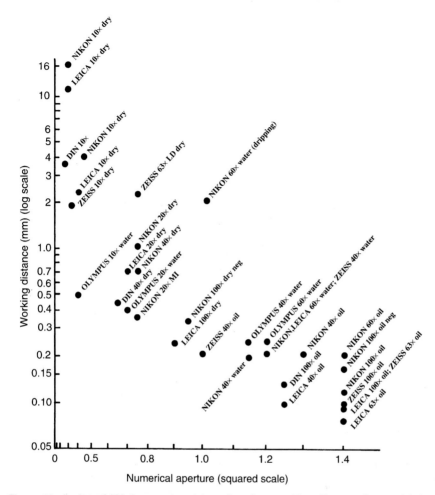

Figure 16. A plot of NA (squared scale) against free working distance (log scale) for a number of commercially available objectives. Note that this table was compiled in 1998 and current offerings from the different makers may be different. The objectives marked 'DIN' are cheap standard lenses from the catalogue of Edmunds Scientific which have been included for orientation: their optical performance should not be equated to that of the named lenses.

Confocal optics do not improve Rayleigh resolution at all, but do improve the Sparrow and FWHM resolution (see ref. 33 for a practical verification for coherent reflection imaging).

7.2.7 How are the lateral and axial resolutions changed by confocal operation?

A good way to think about this is to imagine the two stages of confocal operation as distinct events. The probability of illumination (i.e. the arrival of

Box 1. Resolution criteria

Rayleigh criterion: Two points are said to be resolved if the first minimum (around the Airy disc) of one overlaps the peak of the other; they are then 3.83 optical units apart (see Appendix 3 for a definition of optical units). This criterion is severe, in the sense that the doubleness is obvious to the eye even when the two points are closer than this.

Sparrow criterion: Two points are resolved if the intensity sum of the two has a minimum value between the two peaks. As the two points are brought closer the dip in intensity between the peaks becomes reduced. The Sparrow limit is the situation where the intensity profile between the peaks becomes flat. At this limit, the doubleness is readily visible to the eye: there is an oval rather than circular spot.

FWHM: This measure of resolution is simply the width of the Airy disc at half its maximum height. It is, coincidentally, about the same as the radius of the first minimum.

a photon at a particular point in space) is proportional to the intensity of the point-spread function at that point. The probability of detection (assuming ideal confocal performance) is also proportional to the point-spread function intensity, since this corresponds to the point-spread function of the detector aperture projected back into object space. The resolution (neglecting any effects such as a change in wavelength between illumination and emission) is determined by the product of the two probabilities, which is the square of the intensity of the point-spread function.

What is the effect of squaring the point-spread function? The position of the minima does not change, so the resolution, according to the strict

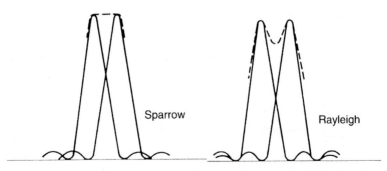

Figure 17. The Sparrow criterion is less stringent than the Rayleigh one: it counts two points as resolved if the intensity profile is flat between the peaks of the two Airy discs formed by incoherent imaging (see text).

Rayleigh definition, does not improve. However, the FWHM improves by a factor of 1.4 (34). So, the lateral resolution becomes:

$$r_{fwhm} = 0.44\lambda/NA \text{ confocal} \tag{1}$$

compared with

$$r_{fwhm} = 0.61\lambda/NA \text{ conventional} \tag{2}$$

For axial resolution the corresponding equations are:

$$Z_{fwhm} = 2\lambda/n(\sin^2\theta) \text{ (13)} \tag{3}$$
$$= 2\lambda/n(\sin^2 \sin^{-1}(NA/n))$$

where n is the refractive index of the medium, and θ is the angle between the outermost ray entering the objective and the optical axis.

At 488 nm, using an oil-immersion objective ($n = 1.515$) of NA 1.4, the values for *Equations 1, 2* and *3* are 0.15, 0.21 and 0.76 μm, respectively.

7.2.8 What can be done about chromatic aberration in confocal imaging?

Severe chromatic effects may be due to the use of the wrong type of objective. Do not use old-style non-compensated objectives such as the Olympus S-Plan series with confocal scan heads designed for modern objectives which give an achromatic intermediate image. The S-Plans were designed to be used with an equal and opposite chromatic aberration in the eyepiece to cancel their own.

Weaker but worrying effects occur with all objective lenses. Researchers who are doing co-localization experiments involving small intracellular particles, bacteria, etc. should test for aberration by collecting complete three-dimensional datasets. It should then be easy to establish whether apparent single positives are real or simply a focus artefact (see Chapter 6).

Strong chromatic aberration is invariably present in objectives over the range from ultraviolet to visible. Ultraviolet confocal apparatus contains lenses designed to counteract this effect. The range of objectives that can be used with such apparatus is strictly limited.

Reflecting objectives of the Schwartzchild type (a combination of a convex mirror and a catadioptric concave mirror, i.e. one with a hole in it) have been suggested as a cure for chromatic problems. They have the additional appeal that spherical aberration can be compensated by adjustment of the mirror separation. However, the point-spread function is badly disturbed by the central occlusion and they are useless for all high-resolution work.

Two-photon imaging (see below) with non-confocal detection provides a cure. Here, it does not matter if the lens is strongly chromatically aberrant, because the resolution is determined entirely by the (monochromatic) illumination and the emission is collected but not focused into an aperture. This will probably become the method of choice for co-localization studies in the future. However, this will demand the use of fluorochromes that can be separated completely by emission windowing alone: FITC and TRITC will not do! (See ref. 35 for a recent determination of multiphoton absorption spectra for a large number of fluorochromes.)

7.3 Multiphoton imaging

Reviews of the theory, apparatus and biological results of multiphoton imaging are available (36–38).

7.3.1 How do multiphoton optics differ from confocal ones?

The laser-scanning microscope of Denk *et al.* (39) depends on the fact that the absorption is proportional to the square of the intensity in the two-photon case, because of the need for two photons to arrive in a short time interval. In the laser microscope, the exciting radiation is focused into a cone, with the result that the absorption falls off very rapidly away from the focus according to an inverse fourth-power rule (see *Figure 18 a*) . The fall-off is even more rapid with three-photon and higher-order processes. Analysis (40) yields the perhaps unexpected result that, provided the illuminated volume is uniformly filled with fluorophores, the total fluorescent emission signal strength is independent of the NA of the objective lens. Unlike confocal microscopy, multiphoton imaging is quite effective with low-magnification lenses of moderate numerical aperture, provided wide-angle detection is used.

The resolution in a two-photon microscope is determined entirely by the restriction of excitation mentioned above. Williams *et al.* (40) show that the lateral resolution is given by $0.37\lambda/n \sin\theta$ and the axial $0.32\lambda/n \sin^2\theta/2$, where λ is the exciting wavelength and $n \sin\theta$ is the NA of the objective lens. These expressions suggest resolution approximately twice as good as a conventional microscope, but this is counterbalanced by the need to use double the wavelength.

In practice, the resolution in two-photon micrographs looks similar to that of conventional epifluorescence, or very slightly inferior if a 1047 nm wavelength is used. A combination of multiphoton excitation and confocal detection is often suggested as a means of improving resolution. This is theoretically justified (41) but is seldom used, because of the loss of signal observed with real fluorescent specimens when the confocal aperture is inserted. The failure of this combination is interesting: it is probably due to scattering of the emitted light in the specimen or chromatic aberration (see ref. 38 for a method for measuring this). Early in the development of two-photon microscopy, the Cornell group found that higher signals could be obtained, with little loss of image quality, by picking off the light coming from the specimen before it entered the scan head, so achieving non-descanned detection. This often increases detection efficiency by at least a factor of three and perhaps more for scattering specimens.

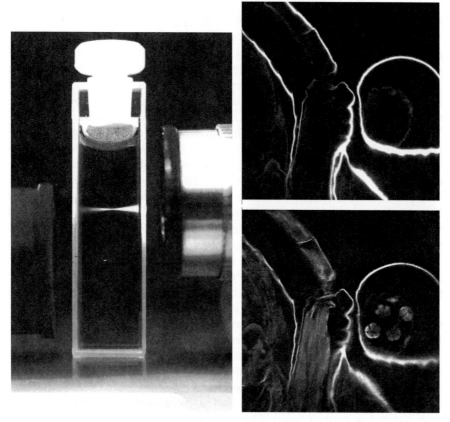

Figure 18. Multiphoton imaging. (Left) A cuvette filled with the fluorescent dye safranin, which normally fluoresces yellow when excited with green light. A beam of green light (543 nm) enters from a lens on the right which focuses it into the cuvette. As expected for ordinary fluorescence, there is a cone of excitation with attenuation because of absorption to the left. By contrast, the lens at lower left focuses a pulsed beam from a mode-locked NdYLF laser at 1046 nm. This beam produces no detectable excitation except at a tiny spot in the centre of the cuvette. Since the excitation requires two photons to arrive within a very short interval, it falls off according to the inverse of the intensity squared, which results in a fourth-power fall-off because of the geometry of the focused beam. (Right) Comparison of a single-photon image (above) and a two-photon image (below) of a waterflea (*Ceriodaphnia* sp.) fixed in ethanol and stained with eosin. The confocal image was obtained with 514 nm excitation: note that little structure is visible in the interior, probably because of intense absorption of the exciting radiation in the exoskeleton. The two-photon image, obtained with 800 nm excitation from a Ti-sapphire laser, shows much internal structure, including muscles and lenses of the compound eye. Field width = 300 μm.

7.3.2 Apparatus for multiphoton imaging

The most-used type of scan head for multiphoton imaging is not very different from that used for confocal imaging. The most important modifications are special coatings to reflect or exclude the infra-red beam used for excitation and provision of a direct high-efficiency path for the non-descanned detection mentioned above. Recently, slit-scanning and multi-spot multiphoton microscopes have been developed, in which cameras are used (42–44).

The laser must have a wavelength approximately twice that of the single-photon absorption peak. Because the signal power is proportional to the square or cube of the excitation power, there is a great advantage in delivering the energy in the form of very short pulses, of the order of tens of femtoseconds. This demands a costly laser of the mode-locked type. Some two-photon images have been obtained with cheaper continuous-wave (CW) lasers but only with very bright specimens and high excitation powers (210 mW compared with the 1–5 mW normally used).

So far, the most commonly-used lasers are titanium-doped sapphire (Ti-sapphire) with a tuning range of 690 nm to approximately 1000 nm and the fixed wavelength neodymium-doped yttrium lanthanum fluoride (Nd-YLF) which is a fixed-wavelength source (1046 nm) which is much more compact than a Ti-sapphire. It is advantageous to use special apparatus to compensate for the pulse-lengthening effects of lenses and coatings in the microscope.

Currently, there is a dispute over the optimum length of pulse. The original discoverers of two-photon imaging (39) specified the shortest attainable pulses (tens of femtoseconds). Assuming no saturation, one would expect the signal to vary with the inverse of pulse duration. With such short pulses (though from a different type of laser) White, in April 1995, was first to demonstrate three-photon imaging (45), and here short pulses are essential, since the signal is proportional to the inverse square of the pulse length. The disadvantage of femtosecond pulses is that they cannot easily be passed through an optical fibre. Convenient and compact two-photon systems have been made with picosecond pulses passed to the microscope through a fibre. Such systems do not yield three-photon images and their advocates argue that three-photon absorption could be a source of damage to the specimen. However, this so-far hypothetical form of damage can be countered by the abundant evidence of single- or two-photon damage due to the use of high intensities of the fundamental (infra-red) wavelength.

7.3.3 Pros and cons of multiphoton versus confocal imaging

While the physical fact that absorption is restricted to the plane of focus in multiphoton imaging (as in *Figure 18*) seems to demonstrate the unequivocal superiority of this mode of microscopy, detail of real biological advantages has been slow to emerge.

On the reduction of light-induced damage, the most convincing result so far is that of Squirell *et al.* (46) in which repeated three-dimensional datasets have been collected from light-sensitive mammalian (hamster) embryos with survival and successful re-implantation and birth. This work was done with a cytoplasmic mitochondrially-localixed fluorochrome (Mitotracker from Molecular Probes). Unfortunately, similar longevity seems not to have been obtained with any nuclear fluorochrome.

On better penetration of the excitation and an improved ability to image at deeper levels (see *Figure 18b* and *c*) the most impressive results have been those of Denk, Svoboda, Just and colleagues (e.g. ref. 47) who have imaged neurons by fluorescence at greater depths than has previously been possible in vertebrate brain tissue.

It has become clear, particularly from the work of Zipfel (36), that multiphoton imaging makes possible types of fluorescence imaging that were completely impossible before. For example, the elastin and collagen fibres deep in intact mammalian skin can be imaged by their autofluorescence. To do this by ordinary single-photon fluorescence is impossible, because the required ultraviolet excitation cannot be made to penetrate deep enough into the tissue. It is hoped that pathologists will take an interest in the remarkable possibilities suggested by this work.

The chief demerits of multiphoton imaging, apart from the high cost of the laser, are as follows. The infra-red beam is hazardous to the retina. It is strongly absorbed by melanin and some other pigments, although the YLF beam seems relatively harmless to chloroplasts. Workers at NIMR, Mill Hill, have found that an infra-red beam can destroy or have a direct triggering effect on motor end-plates (D. Ogden and N. Kiskin, personal communication). Since the two-photon absorption spectra of many dyes are broader than the conventional single-photon ones, there may be difficulties in using the method for fluorescent resonance energy transfer (FRET) experiments.

8. How to choose a fluorescence imaging system

The instrument for a given type of work can be chosen from the key in *Box 2,* which follows the principles of a taxonomic key.

Box 2. Key for choosing fluorescence imaging system

1. Number of fluorochromes requiring simultaneous imaging:
   i. four or more [go to **2**]
   ii. fewer than four [go to **3**]
2. i. Simple system, sub-optimal light economy: *pushbroom spectral imager (camera-based)*
   ii. More complex system, better light economy: *Fourier spectral imager (camera-based)*

Box 2. *Continued*

3. i. Specimen motionless over a period of seconds [go to **4**]

  ii. Specimen moving [go to **6**]

4. i. Image usable without optical sectioning or optical background reduction: *cooled CCD camera*

  ii. Image background needs optical reduction: *TIRF plus cooled CCD or photon-counting single-spot confocal*

  iii. Optical sectioning necessary [go to **5**]

5. i. Specimen punctate or filamentous: *cooled CCD camera plus digital deconvolution* (see Chapter 6)

  ii. Specimens diverse, including those with large volumes of uniform fluorescence, optical sections required for immediate inspection without computation delay: *point-scanning confocal, preferably with analogue and photon-counting mode;. multiphoton point-scanning system*

6. i. Fluorescence level low but well above background. High noise background tolerable, few or no very bright centres in specimen. Optical sectioning unnecessary: *conventional optics plus intensified CCD*

  ii. Optical sectioning necessary as well as high time resolution [go to 7]

7. Full framing required [go to **8**] or not [go to **9**]

8. i. Quality of optical sectioning not critical: *parallel confocal (Nipkow or slit) scanner; multifocal multispot scanner*

  ii. Optical sectioning must approach theoretical optimum with all specimens *high speed confocal resonant galvo scanner; AOD confocal scanner*

9. Line scanning or fast scanning of a selected small area at low pixel number acceptable: *conventional point scanning confocal; AOD; or resonant galvo scanner*

Acknowledgements

I thank Oliver Sedlacek, Wasi Faruqui and Viki Allan for their advice on this chapter.

References

1. Betzig, E. and Chichester, R. J. (1993). *Science*, **262**, 1422.
2. Thompson, N. L. and Lagerholm, B. C. (1997). *Current Opinion in Biotechnology*, **8**, 58.
3. Tokunaga, M., Kitamura, K., Saito, K., Iwane, A. H. and Yanagida, T. (1997). *Biochemical and Biophysical Research Communications*, **235**, 47.

4. Conibear, P. B., Kuhlman, P. A. and Bagshaw, C. R. (1998). *Advances in Experimental Medicine and Biology*, **453**, 15.
5. Ha, T., Enderle, T., Ogletree, D. F., Chelma, D. S., Selvin, P. R. and Weiss, S. (1996). *Proceedings of the National Academy of Sciences of the USA*, **93**, 6264.
6. Inoué, S. and Spring, K. R. (1997). *Video Microscopy*, 2nd edn. Plenum Press, New York.
7. Berland, K., Jacobsen, K. and French, T. (1998). In *Methods in Cell Biology. Video Microscopy* (eds G. Sluder and D. E.Wolf), Vol. 56, pp. 20–45, Academic Press, San Diego, C.A.
8. Feynman, R. P. (1990). *Q.E.D. The Strange Theory of Light and Matter*. Penguin, London.
9. Perrin, F. H. (1966). In *The Theory of the Photographic Process* (ed. T. H. James), 3rd edn, pp. 499–551. Macmillan, New York.
10. Entwistle, A. (1999). *Journal of Microscopy*, **192**, 81.
11. Silver, P. A. (1998). In *Green Fluorescent Protein* (ed. M.Chalfie and S. Kain), pp. 338–340. Wiley-Liss, New York
12. Svoboda, K. and Block, S. M. (1994). *Annual Review of Biophysical and Biomolecular Structure*, **23**, 247.
13. van der Voort, H. T. M. and Brakenhoff, G. J. (1990). *Journal of Microscopy*, **157**, 105
14. White, J. G. (1987). UK Patent Application GB 2 184 321 A.
15. Brelje, T. C., Wessendorf, M. W. and Sorenson, R. L. (1993). *Methods in Cell Biology*, **38**, 97.
16. Gunter, W. D., Erickson, E. F. and Grant, G. R. (1965). *Applied Optics*, **4**, 512.
17. Wu, J-Y. and Cohen, L. B. (1993). In *Fluorescent and Luminescent Probes for Biological Activity* (ed. W. T. Mason), pp. 389–404. Academic Press, London.
18. Trundle, E. (1987) *Television and Video Engineers' Pocket Book*. Heinemann Newnes, Oxford
19. Holst, G. C. (1998). *CCD Arrays, Cameras and Displays*. SPIE Optical Engineering Press. Bellingham, WA, USA.
20. Welford, W. T. and Winston, R. (1989). *High Collection Non-imaging Optics*. Academic Press, New York.
21. Garini, Y., Katzir, N., Cabib, D. and Buckwald, R. A. (1996). In *Fluorescence Imaging Spectroscopy and Microscopy* (ed. X. F. Wang and B. Herman), pp. 87–124. Wiley, New York.
22. Amos, W. B. (1999). PCT Patent application no. PCT/GB98/02236.
23. Foskett, J. K. (1988). *American Journal of Physiology*, **255**, C566.
24. Spring, K. R. (1990). In *Optical Methods in Biology* (eds B. Herman and K. Jacobsen), pp. 513–22. Wiley-Liss, New York.
25. Minsky, M. (1988). *Scanning*, **10**, 128.
26. Amos, W. B. (1991). US Patent 4,997,242.
27. Tsien, R. Y. and Bacskai, B. J. (1995). In *Handbook of Biological Confocal Microscopy* (ed. J. Pawley), 2nd edn, pp. 459–78. Plenum Press, New York.
28. Draaijer, A. and Houpt, P. M. (1988). *Scanning*, **10**, 139.
29. Petran, M., Hadravsky, M., Egger, D. and Galambos, R. (1968). *Journal of the Optical Society of America*, **58**, 661.
30. Tanaami, T., Sugiyama, Y. and Mikuria, K. (1994). *Yokogawa Technical Report English Edition No. 19*, 7.

31. Lichtman, J. W., Sunderland, W. J. and Wilkinson, R. S. (1989). *New Biologist*, **1**, 75.
32. Pawley, J. (ed.) (1995). *Handbook of Biological Confocal Microscopy*, 2nd edn. Plenum Press, New York.
33. Oldenbourg, R., Terada, H., Tiberio, R. and Inoué, S. (1993). *Journal of Microscopy*, **172**, 31.
34. Brakenhoff, G. J., Blom, P. and Barends, P. (1979). *Journal of Microscopy*, **117**, 219.
35. Xu, C., Williams, R. M., Zipfel, W. and Webb, W. W. (1996). *Bioimaging*, **4**, 198.
36. Xu, C., Zipfel, W., Shear, J. B., Williams, R. M. and Webb, W. W. (1996). *Proceedings of the National Academy of Sciences of the USA*, **93**, 10763.
37. Centonze, V. E. and White, J. G. (1989). *Biophysical Journal*, **75**, 2015.
38. Amos, W. B. (1998). *Current Protocols in Cytometry Unit 2.9.* (http://www.wiley.com/cp)
39. Denk, W., Strickler, J. H. and Webb, W. W. (1990). *Science*, **248**, 73.
40. Williams, R. M. Piston, D. W. and Webb, W. W. (1994). *FASEB Journal*, **8**, 804.
41. Higdon, P. D., Toruk, P. and Wilson, T. (1999). *Journal of Microscopy*, **193**, 127.
42. Brakenhoff, G. J., Squier, J., Norris, T., Bliton, A. C., Wade, M. H. and Athey, B. (1995). *Journal of Microscopy*, **181**, 253.
43. Bewersdorff, J., Pick, R. and Hell, S. W. (1998). *Optical Letters*, **23**, 656.
44. Buist, A. H., Muller, M., Squier, J. and Brakenhoff, G. J. (1999). *Journal of Microscopy*, **192**, 217.
45. Wokosin, D. L. Centonze, V. E., Crittenden, S. and White, J. (1997). *Bioimaging*, **4**, 208.
46. Squirrell, J. M., Wokosin, D. L., Centonze, V. E., White, J. G. and Bavister, B. D. (1996). *Molecular Biology of the Cell*, **7**, 644a.
47. Svoboda, K. , Denk, W., Kleinfeld, D. and Tank, D. W. (1997). *Nature*, **385**, 161.
48. Xiao, G. Q. and Kino, G. S. (1987). *Proceedings of SPIE 809: Scanning Imaging Technology* (eds T. Wilson and L. Balk), pp. 107–13, SPIE Press, Bellingham, USA.

5

Fluorescence microscopy of living vertebrate cells

RAINER PEPPERKOK and DAVID SHIMA

1. Introduction

Fluorescence microscopy of living cells has always been a powerful tool for cell biologists. Presently it is experiencing an unexpected renaissance and is on its way to having a place in almost every branch in modern biology from developmental biology to cellular biochemistry. Labelling of organelles or structures in living cells, compared with work in fixed cells, has the obvious advantage that the dynamics of the labelled structures or molecules can be studied. Also, labelling in living cells does not suffer from possible problems induced by the fixation or permeabilization of the cells as is necessary for immunofluorescence or electron microscopy.

The popularity of fluorescence microscopy of living cells is predominantly due to two recent improvements in the technology. The first is the development of new sophisticated imaging hardware. The increased speed of computers available for data handling and image analysis, the availability of single or multiple photon excitation confocal microscopes for the three-dimensional (3D) or even four-dimensional acquisition of data, and the improvements in imaging detectors such as high-resolution cooled slow-scan CCD cameras are all examples of the tools now available which make fluorescence imaging of living cells easier and better than ever before.

The second major improvement in the technology is the development of new fluorescent markers with excellent specificity for the target organelles or molecules. One of the most significant in this respect was the cloning, characterization and continuous improvement of the green fluorescent protein (GFP). GFP and its mutants have the advantage that they are available as cDNA clones and thus can be fused to any gene of interest by simple molecular biology methods. As GFP does not require any co-factors to fluoresce and as it can be easily expressed in cells by standard transfection methods of the respective cDNA, GFP fusion proteins are frequently used as molecule-specific fluorescent markers in living and fixed cells.

In this chapter we first summarize and discuss the hardware essential for fluorescence imaging of living cells. Then we describe methods of *in vivo* labelling of cellular targets by labelling of proteins and micro-injecting them into cells. Next, we focus on the fluorescent labelling of molecules in living cells by using GFP and we discuss the potential pitfalls of this approach. Finally we give two specific examples of fluorescence microscopy on living cells: the study of the breakdown of the Golgi complex during mitosis, and membrane traffic in the secretory pathway.

2. Hardware and software requirements

The hardware and software components of an imaging system for fluorescence microscopy of living cells can be divided into two classes:

(a) those which determine image and data quality, including:

- temporal, spatial and spectral resolution
- the imaging detector sensitivity (signal-to-noise ratio)
- the image analysis hardware and software used to quantify and document the results (see also Chapters 4 and 6);

(b) those with an influence on cell behaviour and thus on the biological readout of an experiment. Cells should be kept as close as possible to physiological conditions, which requires their incubation in appropriate microscope chambers regulating temperature and pH of the culture medium. Also, cells should be exposed to as little excitatory light as possible in order to prevent photobleaching of the fluorophore which results in the formation of cytotoxic radicals. This is best achieved by use of the most sensitive imaging detector and most efficient microscope, with a minimum loss of the emitted fluorescence light *en route* from the labelled cell to the imaging detector.

Unfortunately, some of these components have features which are mutually exclusive. For example, the imaging detectors with the best spatial resolution are not very sensitive. This requires, in order to obtain a reasonable signal above background, exposure of the cells to high amounts of light, which interferes significantly with cell behaviour and function. In contrast, sensitive imaging detectors, such as intensified charge-coupled device (CCD) cameras, require only a fraction of the exposure light compared with high-resolution cameras in order to detect the fluorescent signals above background (see Chapter 4). However, these cameras usually suffer from a very low image resolution and details of labelled structures are very often missed. Thus the ideal microscope set-up and imaging detector for live cell imaging do not exist, in our opinion, and the suitability of the experimental set-up is therefore always dependent on the biological experiments one has in mind.

In the following paragraphs we will discuss the advantages and disadvantages of selected state-of-the-art hardware components which we have successfully used in the past for fluorescence microscopy of living cells.

2.1 Microscopes and associated hardware

The microscope and associated hardware components are the most important components in live-cell imaging as they are used to collect the data. It is no use having good labelling of structures in living cells if the microscope set-up is such that the emitted fluorescence cannot be collected with the appropriate sensitivity or at the speed required to image the biological process of interest. Therefore, the microscope set-up has to be carefully chosen to carry out the experiments in mind, and the final composition may vary from case to case. Although a detailed discussion of individual microscope techniques is beyond the scope of this chapter, we will briefly discuss the considerations to be made for setting up a time-lapse fluorescence microscope for live-cell imaging.

Two major modes of fluorescence microscopy of living cells are currently most widely used for time-lapse microscopy: laser-scanning confocal microscopy (see below) and wide-field 3D sectioning microscopy (Chapter 4). For both cases an inverted microscope is best suited as this allows easy manipulation of cells, such as the addition of drugs or micro-injection of proteins or effectors during the observation.

2.1.1 3D sectioning microscope

As in most applications, both spatial and temporal information is required, so a 3D sectioning fluorescence microscope should be capable of acquiring images automatically, even at different wavelengths. There are many suppliers who provide the necessary equipment for this purpose.

Excitation and emission filter wheels equipped with the appropriate excitation and emission filters, which can be controlled by computer, are available and allow automatic image acquisition of samples stained with different fluorescent markers emitting at different wavelengths. In many applications with living cells, attenuation of the excitation light is also important. In such cases, the excitation filter wheel should also allow the possibility of combining the excitation filters with a selection of spectrally neutral density filters to reduce the excitatory light provided by the microscope's light source in a controlled way.

In epifluorescence mode, in addition to the excitation and emission filters, a dichroic mirror is used for each fluorophore to be detected in order to suppress the stray light of the excitatory light. For the fast and reliable collection of data it is of great importance not to move this part of the microscope during sequential acquisition at different wavelengths, as this generally results in loss of speed or in a non-reproducible distortion of the spatial relationship between the differently labelled structures (pixel shifts). Therefore, a number of custom-made dichroic mirrors are available which are manufactured for the simultaneous acquisition of up to four different colours. Unfortunately, the spectral characteristics of these dichroic mirrors, necessary to separate fluorescence of the different fluorophores properly, are at the price of a significant loss in the emitted fluorescence for each individual colour.

Automated control of the microscope z-focus drive with sub-micrometre resolution should be used for 3D sectioning of the sample. This then allows off-line reconstruction of the 3D distribution of the fluorescently labelled molecules using one of the numerous image deconvolution algorithms available.

When filter wheels are used as described above, image acquisition for different colours has to be sequential. This poses significant problems when fast (<1 sec) processes are to be studied. In such cases structures move significantly during sequential image acquisition and the spatial relationship between the different markers may be distorted. A possible solution to this problem might be to use exposures which are significantly shorter than the kinetics of the process studied. This may, however, not always be possible, as the fluorescence emitted by the cells may be too weak to be properly detected above background at decreased exposure times. Alternatively, colour cameras may be used to simultaneously acquire the fluorescent signals emitted by the markers in use.

2.1.2 Confocal laser-scanning microscopes

Commercially available confocal laser-scanning microscopes clearly overcome some of the limitations of 3D sectioning fluorescence microscopes described above. Simultaneous image acquisition for up to four different colours is possible. Also, the image data acquired at each focal plane only contain a minimum of out-of-focus background fluorescence and 3D image reconstruction does not require time-consuming image deconvolution.

Also, multi-photon excitation laser-scanning microscopes are now becoming available (Chapter 4). Since the light source in such microscopes is a laser, emitting far-red or infra-red laser light pulses (usually with wavelengths of 700–1000 nm), these microscopes appear to be best suited for live-cell imaging, as such laser light is less harmful to living cells than the excitatory light of conventional laser-scanning microscopes (360–647 nm).

However, the light budget of laser-scanning microscopes appears to be poor, as most of the emitted fluorescent light is discarded in order to obtain confocal images. Therefore, weakly labelled structures, detectable with a 3D sectioning fluorescence microscope, may not be detected with a laser-scanning confocal microscope. Also, image acquisition with laser-scanning microscopes is relatively slow (in the range of seconds) and fast processes cannot be imaged unless resolution is greatly compromised.

2.2 Imaging detectors for 3D sectioning microscopy

The camera used to acquire the images in 3D sectioning microscopy is of paramount importance, since it is the device which collects the raw data.

Although there has been considerable progress in camera design, it should be noted that the ideal camera capable of covering all biological applications in live-cell imaging still does not exist. In the following section we will discuss some features of the camera systems most frequently used in live-cell imaging. These considerations should help to decide which camera to attach to a time-lapse microscope with respect to the experiments in mind.

2.2.1 CCD cameras

The image detector in CCD cameras consists of a CCD semi-conductor chip, made up of an array of distinct wells, which can be addressed individually (Chapter 4). The wells have a specific size and capacity (*Table 1*). Each well corresponds to one pixel in the acquired image and thus the size and density of the wells determine the resolution of the camera. In contrast to imaging detectors using image intensifiers (see below) or image tubes, CCD cameras offer an excellent linearity over a broad range of incoming signal, and they show little geometric distortion at the camera edges. Therefore, they are very well suited for quantitative imaging.

Many CCD cameras which are presently commercially available allow the integration of the signals generated by the incoming fluorescent light on the CCD chip before it is read out and the signals digitized. Thus they offer the flexibility of imaging both very bright and extremely faint signals by varying the sample exposure and camera integration times accordingly. As these CCD cameras usually also provide very good image resolution they are widely used for high-quality imaging of living cells. In comparison with video-rate CCD cameras, such on-chip integrating CCD cameras achieve very high detection sensitivities by using extended exposure times. They also offer a feature called 'binning'. Using the binning function, wells are grouped and the corresponding signal added together on the CCD chip. In this way the sensitivity of the camera is increased, by the binning factor, although this is at the price of reduced image resolution. Due to their mode of operation, on-chip integrating CCD cameras mostly cannot be operated at video speed (50 or 60 Hz). The frequency with which these cameras can acquire images depends on the image size (chip size), the integration time needed to acquire a good image, the data readout speed and the rate of data transfer to the computer (see *Table 1* for typical values). It should also be noted that cooling of the CCD chip, usually below 0 °C, is extremely important for the quality of the image obtained with such cameras. This is because a major source of noise generated by the CCD chip is thermal noise, which is also integrated together with the fluorescent signal of interest. Thus the signal above background is not very significantly improved with such cameras by using longer exposure times when they are operated with the CCD chip at room temperature. Cooling the CCD chip down to –20 °C can easily be achieved using Peltier elements combined with air cooling. This is sufficient to reduce the thermal noise such that high-quality imaging of living cells is usually possible.

Table 1. Features of a Princeton Instruments Micromax 1400/1 slow-scan cooled CCD camera

Camera feature	Definition	Typical values
Chip size	Number of wells (pixels) in x,y direction, corresponding to the number of image pixels	1317 × 1035 wells (pixels)
Well size	Size of an individual well which corresponds to one pixel; this determines the image resolution	6.9 μm × 6.9 μm
Well capacity	Number of electrons which can be integrated by a single well before saturation is reached. This sets an upper limit to how many grey values can be distinguished after A/D conversion[a]	45#000 electrons
Quantum efficiency (QE)	Percentage of incoming photons that are converted into an electronic signal	For λ = 400 nm, QE = 5% For λ = 550 nm, QE = 40% For λ = 700 nm, QE = 45%
Dynamic range of A/D[a] conversion	Corresponds to the number of distinct grey values that can be obtained for a fully saturated image	12 bit, equals 4096 distinct grey values
Readout speed	Speed at which the data on the entire chip are read out	1 MHz, equals 1.4 s for the readout of a full-size image
Cooling temperature	Temperature at which the CCD chip is operated	–20°C
Readout noise	Noise generated by readout of the CCD chip	10 electrons

[a]Analogue to digital.

Further parameters which have to be considered when deciding which camera to use for an experimental set-up are the quantum efficiency (QE), well capacity, and the overall geometry of the CCD chip (see *Table 1* and Chapter 4).

The QE of the CCD target is defined as the percentage of incoming photons that are converted by it into a useful electronic signal. The QE is strongly dependent on the wavelength of the incoming light and usually does not exceed 40% for wavelengths in the visible range (400–700 nm) of the spectrum. The QE is very poor below 400 nm for most CCD chips, but several camera suppliers offer specially coated chips which achieve quantum efficiencies of up to 10% in the UV range. Also, 'back-illuminated' cameras now offer up to 80% QE. However, they are more expensive than 'front-illuminated' cameras and have bigger wells (see below), which decreases the image resolution.

The size of the CCD wells determines image resolution. The size of one CCD well corresponds to the physical size of one image pixel. Cameras are available with CCD chips which easily cover the resolution of the microscope

(Chapter 4). CCD chips with bigger wells offer less image resolution depending on the magnification used in the individual experiment.

The capacity of the wells determines the dynamic range. The well capacity determines the amount of incoming signal that can be integrated before the well is physically saturated. When this occurs part of the signal will 'spill over' to the neighbouring wells and will thus distort the image, and the linearity and the resolution of the camera will be lost. Therefore, cameras which are manufactured to operate within a high dynamic range (e.g. 16 bit, which corresponds to 65536 distinguishable grey values), have bigger wells to overcome this problem. It should also be kept in mind when deciding which camera to use that, although some camera systems use a 12 or 16 bit data digitization protocol, the well capacity of the CCD chip used may be too low to give a true 12 or 16 bit capacity.

2.2.2 Intensified CCD cameras

As described above, on-chip integrating cooled CCD cameras achieve very good sensitivities by using longer sample exposure times and integration of the weak incoming signals. However, many biological processes occur in the range of a few seconds or even sub-seconds. In order to image those processes, cameras that are more sensitive and faster than the on-chip integrating CCD cameras are needed, such as intensified CCD cameras. These typically consist of a type IV image intensifier coupled to a CCD chip via a lens or optical fibres. These cameras are orders of magnitudes more sensitive than normal CCD cameras when comparing identical exposure times. Therefore, with these cameras, the exposure light can be attenuated to a minimum, which is less harmful to cells and causes less photobleaching. Also, images can be acquired at, or close to video rate. However, the image resolution, dynamic range and geometric distortion at the edges of these cameras are determined by the image intensifier, whose performance with respect to these parameters is usually considerably worse than that of the respective non-intensified CCD cameras. Thus intensified CCD cameras are used in experiments where less spatial resolution but high temporal resolution is required (e.g. calcium imaging or vesicular transport) and when the fluorescent signals are very weak.

2.2.3 Colour cameras

So far we have discussed black-and-white cameras capable of imaging single wavelengths at a time. In order to carry out multi-colour imaging with these cameras, the images at different wavelengths have to be acquired sequentially by changing the excitation and emission filters using filter wheels (see above). Therefore, to acquire a multi-colour image requires considerably more time than imaging a single colour, and hence this may be insufficient to image rapid processes. Also, when images at different colours are acquired sequentially, the objects of interest may move during image acquisition and the spatial relationship between the differently labelled objects (with respect to their colour) will be distorted.

In order to overcome these problems, a confocal microscope may be used (see above). Alternatively, three colour on-chip integrating CCD cameras (with red, green and blue detectors) may be used. In combination with the appropriate markers (they should emit red, green and blue light), they allow the simultaneous visualization of up to three different markers at a speed greater than that achieved with confocal microscopes. However, some fluorescent markers that can be used very efficiently to label cellular structures (different GFP mutants, for example) do not emit fluorescent light that can be well separated into the spectral components of a colour camera. Thus, quantitative association of the detected signals with the individual markers in use is impossible.

2.3 Temperature and environmental control

One of the most crucial aspects in fluorescence microscopy of living cells is the environmental conditions of the cells during imaging. Cells should be kept as close as possible to physiological conditions. The key parameters in this respect are temperature and the pH of the incubation medium. A number of commercially available solutions exist to this problem and are available from most microscope manufacturers. In our experience, however, the best results are achieved by enclosing the entire microscope in a cheap home-made Perspex box. The temperature is easily controlled by this means and, once equilibrated, remains stable for hours. An additional advantage of this approach is that it makes the system more stable physically and helps the sample to stay in the focal plane. This is important for experiments that require continuous imaging of the same cells for several hours, as is the case for studies on cell-cycle regulation or apoptosis. The pH of the incubation medium may be controlled by using carbonate-free medium buffered with 20 mM Hepes, pH 7.3. Alternatively, the Perspex box may be flooded with CO_2 when carbonate-containing medium is used.

2.4 Data analysis, image processing and data presentation

In recent years computer systems have become powerful enough even to perform 3D reconstructions involving time-consuming deconvolution algorithms at a reasonable speed. This has considerably increased the number of powerful and easy-to-use software packages commercially available. One of the most popular public-domain software packages for scientific image analysis and documentation is the NIH image program (available from: http://rsb.info. nih.gov/nih-image/). It allows the generation of simple macros, without the user having a background in computer programming, and thus allows the automated acquisition and evaluation of image data. It covers most basic features in image analysis, including the control of microscope peripherals. It is also available as a source code, which allows users to implement their own extensions covering more sophisticated demands.

In general, the software package of choice for the more sophisticated analysis and documentation of time-lapse image data should have the possibilities for automatically analysing geometric and densitometric parameters on multiple colour images in four dimensions (time and space). Also, tracking of individual objects and visualization of their trajectories should be possible. An effective way of presenting data of live cell experiments is to convert the individual images into Quicktime movies which can be easily placed on to a World Wide Web server from where they can be visualized from the outside world.

3. Fluorescent labelling of molecules and organelles in living cells

A number of approaches for fluorescently labelling cytoplasmic and nuclear structures or organelles in living cells are currently available. They vary considerably in terms of ease of use, how specifically they label the target molecules, and in how much they interfere with cellular function.

3.1 Cell-permeable markers
A simple approach to label cellular target is to use commercially available fluorescent cell-permeable compounds. They are added to the cell culture medium to label the respective cellular structures. However, some of these also cross-react with structures different from the intended targets and therefore it is important to perform a stringent specificity test before such markers are used in real experiments. A short list of such compounds and their cellular targets is shown in *Table 2*. As they have been extensively described in the literature, and as the manufacturers supply labelling protocols, we will not concentrate on them further here.

3.2 Labelling by micro-injection of fluorescently labelled proteins or antibodies
Direct micro-injection of fluorescently labelled molecules into cells is a universal and efficient way of labelling cellular targets. Almost any material that is fluorescently labelled and passes through the injection capillary, may be introduced into cells in this way. In order to avoid interference with cellular function or non-specific cross-reactions, the fluorescent markers should be characterized *in vitro* before they are introduced into cells. At least two points have to be considered in order to optimize labelling by micro-injection:

(a) Molecules are usually diluted 10- to 20-fold in the cell upon micro-injection, so the concentration of markers in the injection pipette will

Table 2. Selection of cell-permeable fluorescent markers[a]

Cellular target	Marker	Comments
Plasma membrane	con A	is slowly internalized
Nucleus	SYTO dyes	exist for several excitation/emission wavelengths
Endoplasmic reticulum	$DiOC_5$, $DiOC_6$	also accumulate in mitochondria
Golgi	NBD ceramide	moves to the plasma membrane at 37°C
Secretory vesicles	acridine orange	accumulates in any acidic organelle
Endosomes	rhodamine–transferrin	labels early endosomes, and is recycled to the plasma membrane
Endosomes	fluorescein–dextrans as fluid-phase markers	move through the entire endocytic pathway
Lysosomes and acidic organelles	lysotrackers	
Mitochondria	rhodamine 123	

Abbreviations: $DiOC_5$, 3,3'-dipentyloxacarbocyanine iodide; $DiOC_6$, 3,3'-dihexyloxacarbocyanine iodide; NBD, 7-nitrobenz-2-oxa-1,3-diazole.

[a]This list of markers is far from complete. All can be obtained from Molecular Probes (http://www. probes.com), which offers a whole range of fluorescent markers for cellular organelles and small regulatory molecules like calcium.

always have to be higher than the optimal concentration used for *in situ* experiments such as immunofluorescence using labelled antibodies. Equally, micro-injecting too much fluorescent material into the cells may result in a high background due to the molecules that have not bound to their cellular target.

(b) If the labelled markers are proteins with a cellular counterpart, it is of great importance that the labelled marker behaves identically to the endogenous molecules. If labelled antibodies are micro-injected one should test to make sure that they do not inhibit cell function.

It is therefore most useful to micro-inject the fluorescent markers over a whole range of different concentrations in order to achieve optimal labelling conditions which do not interfere with cell function. Furthermore, it should be noted here that micro-injection also offers the advantage that combinations of markers can be introduced simultaneously into cells.

3.2.1 Fluorescent labelling of proteins

Fluorescently tagged antibodies or proteins have been the most frequently used markers for *in vivo* labelling experiments. The antibodies or proteins used should be purified and highly active, since contaminating or inactive molecules in the antibody or protein solution will also be fluorescently labelled and thus cause non-specific background fluorescence when micro-injected into cells.

Several bright and photostable fluorescent dye derivatives are available that react with free amine or sulfhydryl groups as found for example in lysines or cysteines, respectively (see *Table 3* for selected examples). Typical dye derivatives used for labelling amines or sulfhydryls are the isothiocyanates and carboxy-succinimidyl esters or iodoacetamidos and maleimides, respectively. When deciding which dye derivative to use for labelling a particular protein with high efficiency, it is important to know whether the specific residue is accessible within the protein. Amine-containing side chains are usually on the surface of proteins and thus are very easy to derivatize. In contrast, the sufhydryl groups in cysteines are usually in the core of proteins and are thus often inaccessible. Although there are methods to circumvent this problem, labelling of these groups could interfere with their function in establishing cysteine disulfide bridges, which are key features in stabilizing protein structure and conformation. As interference of the labelling procedure with protein function is a general and often unpredictable problem, functionality tests prior to *in vivo* use are essential. Below we describe two protocols that we routinely use in our laboratories for labelling antibodies used for labelling cellular targets *in vivo*. They may also be easily adapted for labelling other proteins. A more detailed discussion of fluorescent dye derivatives and their applications to the labelling of *in vivo* markers can be found in ref. 1.

The method outlined in *Protocol 1* aims at the fluorescent labelling of amine groups (present, for example, in lysines) with the rhodamine derivative tetramethylrhodamine 5-isothiocyanate (TRITC; excitation, 546 nm; emission, 579 nm). Following this protocol a labelling efficiency of 2–4 moles of dye per mole of antibody is usually achieved.

Table 3. Amine and sulfhydryl group-specific dye derivatives and their characteristics

Derivative	Ex./Em. (nm)	Specificity	Solvent
Fluorescein isothiocyanate (FITC)	494/520	amine	DMSO
5-Carboxyfluorescein, succinimidyl ester	491/518	amine	DMSO
Fluorescein 5-maleimide	495/520	sulfhydryl	DMF
Alexa488 succinimidyl ester	495/519	amine	water
Alexa488 maleimide	495/519	sulfhydryl	water
Tetramethylrhodamine 5-isothiocyanate (TRITC)	544/570	amine	DMSO
5-Carboxytetramethyl-rhodamine succinimidyl ester	546/579	amine	DMSO
Tetramethylrhodamine 5-iodoacetamide	555/580	sulfhydryl	TES pH 7
Cy3 succinimidyl ester*	550/570	amine	water
Cy5 succinimidyl ester*	649/670	amine	water

All dye derivatives listed in the table and many others with variations in excitation (Ex.) and emission (Em.) maxima are available from Molecular Probes (http://www.probes.com/) except the asterisked ones, which are available from Amersham Pharmacia Biotech (http://www.apbiotech.com/)

Protocol 1.   Labelling antibodies with tetramethylrhodamine
5-isothiocyanate (TRITC)

Equipment and reagents

- 0.1 M sodium carbonate buffer, pH 8.5[a]
- TRITC (Pierce)
- Dimethyl sulfoxide (DMSO)
- Gel-filtration column containing Sephadex 25 (Pharmacia) matrix, or similar

- Glycine, pH 8.5
- Separation buffer, e.g. micro-injection buffer (48 mM K_2HPO_4, 4.5 mM KH_2PO_4, 14 mM NaH_2PO_4, pH 7.2)
- Spectrophotometer

Method

1. Dissolve the antibodies to be labelled in 0.1 M sodium carbonate buffer pH 8.5.[a] The protein concentration should be at least 1 mg/ml.

2. Dissolve TRITC (Pierce) in dry DMSO at a concentration of 0.5 mg/ml. This solution should always be made up freshly, as breakdown of the thiocyanate group over time may drastically decrease the coupling efficiency. Protect the solution from exposure to day light.

3. Slowly add 50μl of TRITC solution per milliliter of protein solution. Mix gently.

4. React for at least 8 h at 4°C in the dark. Occasionally mix the sample gently.

5. Stop the reaction by adding glycine at pH 8.5 (final concentration 1 mM). Incubate for 2 h at 4°C to block remaining free TRITC, which has not reacted with the antibodies.

6. Separate labelled antibodies from free TRITC by gelfiltration using a Sephadex 25 (Pharmacia) matrix, or similar. To obtain optimum separation the column size should be at least 15 times the sample volume. As separation buffer we usually use micro-injection buffer (48 mM K_2HPO_4, 4.5 mMKH_2PO_4, 14 mM NaH_2PO_4, pH 7.2). As labelled proteins often stick to the gel filtration support, the column should not be reused. Alternatively, the sample may be dialysed against an appropriate buffer for at least 24 h in the dark.

7. Measure the OD of the fractions containing the labelled protein at 280 and 546 nm. The ratio of OD_{546}/OD_{280} should be between 0.15 and 0.5, corresponding to approximately two to six dye molecules per antibody molecule.

[a] Similar buffers within the pH range 8.0–9.0 may be used. However, they must not contain amines (as is the case with Tris buffer) which would compete with the amines to be labelled within the protein resulting in a very limited labelling efficiency.

Protocol 2 outlines the procedure for fluorescent labelling of sufhydryl groups (only present in cysteines) with the fluorescein derivative fluorescein 5-maleimide (excitation, 495 nm; emission, 520 nm).

Protocol 2. Labelling proteins with fluorescein 5-maleimide

Equipment and reagents

- 20 mM sodium phosphate, 0.15 M NaCl, pH 7.2[a]
- Spectrophotometer
- Stock solution of 20mM fluorescein 5-maleimide in dimethyl formamide
- Gel-filtration column (see *Protocol 1*)

Method

1. Dissolve the protein to be labelled at a concentration of at least 1 mg/ml in 20 mM sodium phosphate, 0.15 M NaCl, pH 7.2.[a]
2. Add the dye solution in a 10- to 25-fold molar excess over the protein to be labelled.
3. Incubate for 2 h at room temperature or 8 h at 4°C in the dark.
4. Separate free dye from labelled protein by gel filtration (see *Protocol 1*).
5. Measure the OD at 495 and 280 nm. An OD_{495}/OD_{280} ratio between 0.2 and 0.5 is the optimum for an intense labelling without over-modification of the antibody, which results in its inactivation and thus non-specific background.

[a] Similar buffers within the pH range 6.5–7.5 may be used as long as they do not contain sulfhydryls which would compete with the sulfhydryls of the protein to be labelled, inhibiting fluorescent conjugation.

Protocols 1 and *2* are examples of how to modify amine or sulfhydryl groups. They may be easily adapted to the conjugation of proteins with similar dye derivatives based on isothiocyanates or maleides.

Labelled antibodies should be divided into small aliquots suitable for micro-injection (10 μl), rapidly frozen in liquid nitrogen and stored at –80°C in the dark. Storage for >6 months is not recommended as this often results in the loss of *in vivo* labelling efficiency. Individual aliquots should only be used once for injection as repeated freezing and thawing often results in either the inactivation of the antibodies or their aggregation, which makes injection difficult and causes a disruptive background in the cells.

A variable fraction of the antibodies may have lost their activity due to over-modification by the labelling procedure. As this fraction of antibody is also micro-injected into cells it will cause a cytoplasmic background as it does not bind to its antigen. In order to remove this fraction of inactive antibodies, the labelled antibodies may be subjected to a further round of affinity purification using immobilized antigen. Alternatively, over-modified antibodies may be removed by ion-exchange chromatography.

3.2.2 Micro-injection of fluorescently labelled markers

A detailed discussion of suitable micro-injection equipment and technical aspects is beyond the scope of this chapter and is described in ref. 2.

For *in vivo* labelling experiments we use the same inverted fluorescence microscope and objective (preferably phase-contrast objective) which we use for the time-lapse analysis of living cells. This has the advantage that only a few cells have to be injected, which can then be easily relocated for the subsequent analysis by time-lapse microscopy.

The glass micropipettes used for penetrating cells and delivery of the sample are available from Eppendorf (Eppendorf, Germany). The manipulators and the micro-injector, controlling the pressure applied to the micro-injection pipette, used in our laboratories, are also from Eppendorf.

For micro-injection and time-lapse microscopy, cells are plated on 35 mm dishes which have been engineered with no. 1.5 mm coverglass bottoms (MatTek Corp. or Intracel), allowing the use of high numerical aperture, short working-distance objectives. During injection and time-lapse microscopy, cells are incubated in tissue culture media without Phenol Red and only low concentrations of serum (1%) to minimize background fluorescence from the culture medium. To keep the pH stable, carbonate-free culture medium (Gibco) buffered with 20 mM Hepes pH 7.4 is used. The temperature is kept at 37 °C by enclosing the entire microscope and micro-injection manipulators in a temperature-controlled Perspex box (see above).

Protocol 3. Micro-injection of labelled markers into living cells

Equipment and reagents
- Micropipette with Eppendorf GelLoader tips (Eppendorf, Germany)
- Microcentrifuge
- Eppendorf micromanipulator (cat. no. 5171) and microinjector (cat. no. 5246)

Method

1. Spin the sample to be injected at 10 000 r.p.m. in a microcentrifuge at 4 °C to remove any aggregates, which could block the micro-injection capillary. Take the supernatant and transfer it into a new micro-centrifuge tube.

2. Load the sample (0.5 μl is more than sufficient) into the micropipette through the open rear end using an Eppendorf GelLoader tip.

3. Insert the loaded micropipette into the pipette holder of the micro-injector and lower the pipette until it touches the culture medium. Carefully centre the micropipette tip in the field of view and lower it carefully until it touches the cell to be injected (a white spot appears at the point where the pipette touches the cell).

4. Set the limit of the Eppendorf manipulator to define the cell surface as a reference point.

5. Execute the micro-injection.[a]

6. The same limit may be used to micro-inject nearby cells. They are injected by approaching them at a safe distance above the cells to be injected, with the micro-pipette tip placed above the point of injection.

[a] For injections into the nucleus, the micropipette tip should be placed above the centre of the nucleus. For injections into the cytoplasm, it should be placed as close as possible to the nucleus. Typical injection times, defined by the time the pipette stays inside the cell, vary between 0.1 and 0.5 sec. The pressure applied to the injection pipette varies, depending on the sample and cells to be injected. However, for samples which do not tend to block the pipette tip, pressure values between 50 and 200 hPa are generally suitable for cells with an average size around 30 μm. Typical volumes injected vary between 0.1 and 0.5 pl depending on the place of injection (nucleus/cytoplasm). This corresponds to dilution of the sample in the cell, compared with the concentration in the pipette, by a factor of 1–20.

If micro-injection is carried out carefully, labelled cells can usually be used for time-lapse analysis immediately after injection, provided that labelling of the cellular target by the micro-injected fluorescent marker is fast enough.

4. Using GFP and its variants in live-cell fluorescence microscopy

4.1 Introduction

The advent of GFP has provided a remarkable opportunity to many areas of biological research; particularly live-cell fluorescence microscopy. Moreover, since its initial experimental application 5–6 years ago (3), intensive mutagenesis of the primary sequence has expanded GFP's utility by altering the spectral and biochemical properties of the fluorescent protein (4). GFP and its variants have now been successfully used as specific, multi-colour fluorescent, subcellular probes (5), reporters for gene expression studies, *in vitro* and *in vivo* vital tracers for following cell fate within complex tissues (6), tools for analysis of protein dynamics (7) and adaptable sensors of alterations in the extracellular and intracellular micro-environment (8,9).

4.2 Practical considerations

Selection of the GFP variant with the appropriate spectral properties will allow live-cell visualization over a range of emission wavelengths and provides the potential for simultaneous analysis of multiple GFP chimeras in a single cell (5). *Table 4* lists the most efficient GFP mutants presently available and their spectral characteristics. With this collection of mutants, up to three different GFP chimeras, using enhanced blue fluorescent protein (EBFP), cyan fluorescent protein (CFP) and yellow fluorescent protein (YFP), could be simultaneously visualized (5). However, only a few biological applications visualizing more than one GFP chimera in living cells have been described so far. This is because, although these GFP mutants can be spectrally distinguished, they still contain a significant amount of spectral overlap, requiring particular excitation light sources (5) and combinations of emission filters for their efficient spectral separation in co-localization experiments. As this can only be achieved by discarding considerable amounts (about 50% or more) of the fluorescence emitted by each GFP mutant, the detection sensitivity of the imaging system is greatly reduced. Although this may be partly compensated for by using highly sensitive imaging detectors or extended exposure times in fixed samples, such reduced detection of emitted fluorescence has limiting consequences for live-cell imaging, which requires the cells to be exposed to a minimum amount of fluorescent light in order to keep them biologically active.

The BFP and CFP mutants (see *Table 4*) have excitation maxima at 380 and 433 nm, respectively. Repeated exposure of cells to these short excitation wavelengths may cause more damage to living cells than the excitation at 513 nm that is necessary for the YFP mutant. Thus, although possible, the simultaneous use of BFP, CFP and YFP still has its limitations for live-cell imaging which cannot be compensated for simply by the use of expensive equipment. An alternative, and more efficient, method to co-localize GFP chimeras quantitatively in living cells involves fluorescence lifetime imaging (10). Up to three different GFP chimeras (ECFP, YFP-10C and YFP5; see also *Table 4*) can be distinguished by this method without the need to discard any fluorescence emitted by the GFPs, since spectral separation is not re-

Table 4. GFP mutants and their characteristics

GFP mutant	Excitation (nm)	Emission (nm)	Lifetime $\tau_\phi / \tau_m{}^a$ (ns)	Reference/ source
wt GFP	395/475	508	ND	(4)
EBFP	380	440	ND	Clontech [b]
ECFP	433 (453)	475 (501)	1.32/2.23	Clontech
MmGFP5	473	507	2.42/2.68	(6)
S65T	489	511	2.57/2.59	(4)
EGFP	488	507	2.85/2.88	Clontech
d2EGFP [c]	488	507	ND	Clontech
EYFP	513	527	2.85/2.88	Clontech
YFP5	514	531	3.69/3.60	(10)

Abbreviations: EBFP, enhanced blue fluorescent protein; ECFP, enhanced cyan fluorescent protein; EGFP, enhanced green fluorescent protein; EYFP, enhanced yellow fluorescent protein; GFP, green fluorescent protein; ND, not determined; wt, wild type; YFP, yellow fluorescent protein.
[a] Phase and modulation fluorescence lifetimes, as determined by Pepperkok *et al.* (10).
[b] More information on available GFP expression vectors with multiple cloning sites is available at http://www.gfp.clontech.com/.
[c] This GFP mutant contains a C-terminal PEST sequence that targets the protein for degradation and thus results in its rapid turnover (approximately 2 h).

quired. However, the method is still quite expensive and commercial distribution of the technology is limited.

Alternative approaches to the use of multiple GFP mutants in order to co-visualize more than one molecule in living cells is the combined use of a GFP chimera with cell-permeable fluorescent organelle markers (see above) or micro-injection of fluorescently labelled marker proteins (see above).

Besides the spectral properties of the GFP variants, there are a number of other considerations and, often, limitations (unfortunately more difficult to find outlined in the primary literature or company catalogue), which should be researched carefully before determining suitability for a specific application. It is of great importance to test that the GFP chimera introduced into living cells mimics the properties of the parental protein. Unlike Myc, HA or FLAG tags (16), which are usually less than 20 amino acids long, GFP is >200 amino acids in length and thus cannot be assumed to behave as a neutral addition to the protein of interest.

Moreover, the effect of ectopic expression of chimeras should be assessed for adverse effects on cell behaviour. For example, over-expression of the GFP chimera may compete with the endogenous counterpart for the binding of cellular factors and thus inhibit cell function. Similarly, expression of too much GFP in cells may become toxic owing to the increased appearance of oxidative by-products generated during the folding of GFP after synthesis. These oxidative by-products may be counteracted by adding scavengers like Oxyrase (Oxyrase Inc.), ascorbic acid or glucose oxidase to the culture medium during live-cell imaging.

The fluorescent light emitted by most of the GFP mutants listed in *Table 4* is strongly pH sensitive. Although this pH sensitivity of the GFP mutants has in some cases been successfully exploited to determine the pH in the lumen of the membrane organelles in the secretory pathway (8,9) it has direct implications for their use in live-cell imaging. For example, quantitative studies with GFP chimeras which involve significant pH changes (e.g. transport between the membrane-enclosed organelles in the endocytic or degradative pathway) are impossible when measurements are based on the intensity of the fluorescence emitted by the GFPs.

4.3 Creating GFP chimeras

In addition to an impressive history of free exchange between laboratories, the cDNAs for GFP and a multitude of variants are now becoming widely available commercially (e.g. from Clontech), often in convenient expression vectors and occasionally with optimized codon usage to increase expression efficiency. Thus far, successful fluorescent chimeras with the GFPs have consisted of fusions to the N- or C-terminus, rather than within a central region of the protein of interest. More detailed information about creating GFP fusion proteins is beyond the scope of this chapter, but has been discussed elsewhere in greater detail (11).

4.4 Considerations for stable and transient expression of GFP chimeras in cells

Several options exist for introducing GFP chimeras into cells. First, the choice of transient versus stable expression will depend upon the specific application. For example, to observe GFP-tagged mitotic cells we have opted to create stable cell lines. In this way, we both increase the chances of obtaining mitotic cells (<5% of the population) expressing the fluorescent tag, and we create a fairly uniform, homogeneous level of expression because our stable cell lines are of clonal origin. An added benefit of creating a stable cell line is the ease of characterization of GFP chimeras, in terms of their cellular localization by electron microscopy and/or biochemistry.

General strategies for creating stable cell lines include the creation of clonal lines or the use of mixed clones obtained simply by pooling all antibiotic-resistant clones following transfection. Enhanced levels of transgene expression from stable cell lines have been achieved by treatment with fairly non-specific transcriptional activators such as butyrate (see, for example, ref. 12) or by selection of an enriched fluorescent population by using a fluorescence-activated cell sorter (FACS).

Protocol 4. Expression of GFP chimeras by micro-injection of plasmid DNA[a]

The loading of sample into the micropipettes, mechanical manipulations, penetration of cells and delivery of sample are performed as described above (*Protocol 3*). However, to obtain rapid and reproducible expression of the GFP chimeras, cells should be kept at 37 °C during micro-injection using the equipment for live-cell imaging (see above). Also, the UV fraction of the white light from the microscope lamp used to illuminate the cells should be removed by introducing a green or red filter into the light path between the light source and the cells.

Equipment and reagents

- 35 mm Petri dishes with 1.5 mm cover-glass bottoms (MatTek Corp. or Intracel)
- Standard culture medium, pre-warmed to 37 °C and pre-equilibrated
- Standard culture medium, except containing a low (0.85 g/l) carbonate (Gibco), and buffered with 30 mM Hepes pH 7
- Diamond pen

- Micro-injection equipment (see *Protocol 3*)
- Plasmid DNA for micro-injection, carefully purified by standard protocols and stored in small aliquots at high concentrations (>1 μg/μl in a suitable buffer, e.g. TE) at –20 °C[b] Before use, dilute in syringe-filtered H_2O.[c]
- CO_2 incubator

Method

1. Grow cells on 35 mm dishes, with 1.5 mm cover-glass bottoms.
2. Mark the bottom of the cell dish with a diamond pen. This helps to relocate the cells quickly after micro-injection.

3. For the observation and manipulation of cells in the absence of CO_2, replace the standard culture medium with growth medium containing a low concentration of carbonate (Gibco) and buffer it with 30 mM Hepes pH 7. Otherwise, the medium should contain all the usual supplements used in the standard culture medium.

4. Micro-inject the plasmid DNA directly into the nucleus of the cells (see *Protocol 3*).

5. Replace the culture medium for micro-injection with pre-warmed and pre-equilibrated standard culture medium and incubate the cells at 37°C in a CO_2 incubator.

[a] Following this protocol for DNA micro-injection into cells we obtain visible expression of GFP chimeras in as little as 30 min after micro-injection in approximately 50–70% of injected cells.
[b] Experience in our laboratories has shown that storage of the DNA at 4°C and at low concentrations in aqueous buffer (or H_2O) decreases the expression efficiency and kinetics significantly.
[c] Typical DNA concentrations that are required for strong and rapid expression vary between 0.02 and 0.1 μg/μl (these are the concentrations in the micropipette) depending on the DNA vector used.

However, many live-cell fluorescence applications of GFP require the microscopic observation of small cell numbers, so transient transfection strategies are likely to be sufficient. In addition to the standard transfection procedures, such as electroporation, calcium phosphate or cationic lipid-based gene transfer (13), we often introduce plasmids encoding GFP chimeras directly into the nucleus using micro-injection (see *Protocol 4*). This procedure yields rapid, visible fluorescent protein expression, often within 0.5–1 h, and is particularly useful for quick analysis of a large number of different GFP chimeras.

5. Examples of imaging fluorescent molecules in living cells

In the following sections we describe the considerations and procedural set-up for two distinct types of live-cell imaging. These examples were chosen to highlight the possibilities and pitfalls associated with more demanding live-cell fluorescence microscopy. They include the application of an EGFP chimera (NAGFP; 12) to the analysis of Golgi-complex behaviour during mitosis, and the use of an EGFP-tagged secretory marker (14) together with Cy3-labelled Fab fragments of a monoclonal antibody for double-labelling analysis of membrane dynamics and transport kinetics in the early secretory pathway.

5.1 Reagents and microscopy set-up

We prefer to culture cells on 35 mm dishes engineered with 1.5 mm coverglass bottoms (MatTek Corp. or Intracel). This allows the use of high numerical aperture objectives which require immersion oil for their operation and which have working distances shorter than the thickness of standard plastic tissue culture dishes. In order to avoid disturbing background fluorescence, the culture medium used during the observation of the cells should be free of Phenol Red, antibiotics and fetal calf serum. For visualization in the absence of CO_2, the medium should be buffered with 30 mM Hepes pH 7.4. In order to reduce phototoxicity due to free radicals forming by photobleaching of the fluorophores, ascorbic acid (1 μg/ml) or Oxyrase (30 mM) may be added to the medium as a free-radical scavenger. For long-term observations (>2 h) the culture medium should be covered with mineral oil (Sigma) to avoid the culture medium drying out.

For single-label time-lapse microscopy we use an inverted Zeiss fluorescence microscope equipped with a Sutter filter wheel/shutter for the controlled illumination of cells. Images are acquired with a Princeton Instruments cooled CCD camera (Micromax 1400/1) controlled by the IPLab software package (Signal Analytics, Fairfax, VA, USA) and allows the acquisition of up to four images (600 × 500 pixels) per second. The entire set-up is enclosed in a Perspex box which is temperature-controlled via a heating fan and electronic control unit (plans describing the design of this temperature-controlled box can be obtained from the Mechanical and Electronic Workshops at EMBL, Heidelberg, Germany).

For double-labelling experiments we prefer to use a Zeiss LSM510 confocal microscope. As imaging with two or three colours can be performed simultaneously with this microscope, the spatial relationship between the different fluorescent markers can be imaged with confocal resolution. However, two images at the most can be acquired per second with this microscope, which sets an upper limit for the temporal resolution.

5.2 Mitosis

Mitotic cells are one of the more challenging subjects for time-lapse microscopy. Compared to interphase cells, they have several properties which make them difficult optical specimens:

(a) They are usually spherical or ovoid, so the image within the region of focus represents only a fraction (usually <10%) of the total cell depth in each optical field. This cell shape also leads to increased light scatter, significantly decreasing the efficiency of light detection. Thus, there is more non-specific background fluorescence in mitotic cells, and as a result, a lower signal-to-noise ratio.

(b) They are sensitive to photodamage, so must be visualized with limited light exposure and a highly sensitive detection system, otherwise cells are likely to arrest in early mitosis, or undergo aberrant mitoses.

(c) At least in mammalian cell culture, mitotic cells represent only a small fraction (<5% in a random cycling population) of the total cell number, so careful visual selection of the appropriate cell(s) is often necessary. The use of specific drugs or mechanical methods to increase the population of mitotic cells helps to alleviate this problem (12).

It is currently difficult to overcome all the pitfalls associated with live imaging of mitotic cells using a single detection system. For example, imaging a single section of a mitotic cell with a cooled slow-scan CCD camera provides sufficient sensitivity to detail the behaviour of the Golgi apparatus with fairly high temporal resolution throughout all phases of mitosis (15). One major disadvantage of this imaging method is the high amount of out-of-focus signal in the acquired images. Also, since mitotic cells are usually thicker than interphase cells, imaging only one section means that useful information in the z-axis of the cells is lacking. A solution to this problem is to upgrade from standard fluorescence microscopy to a 3D sectioning fluorescence microscope approach (see section 2.1.1). Such a microscope system has been successfully used to document the behaviour of microtubules in the mitotic spindle; however, it has been limited to temporal resolution on the order of minutes rather than seconds.

An alternative imaging approach is obtaining 3D mitotic images from a z-series using a laser-scanning confocal microscope. This method has the capacity to provide comprehensive information through the cell depth during mitosis and, unlike the CCD camera-acquired images, does not rely upon post-acquisition image restoration procedures to remove out-of-focus inform-ation. However, because the majority of light is discarded during confocal imaging, this technique is significantly less sensitive than one using a 3D sectioning microscope (see section 2.1.1). With the current instrumentation we use in our laboratories (Zeiss LSM 510) we have obtained a maximum of 10–15 optical slices (1 μm apart) through the depth of mitotic cells, at 30 sec to 1 min intervals over a 90 min period (see *Figure 1*).

Although these data give some useful insight into the dynamics of Golgi membranes during mitosis, a more detailed analysis of the complex dynamics associated with mitosis will require overcoming the current technical limita-tions for obtaining images with both high temporal and spatial resolution.

5.3 Visualizing membrane transport in the secretory pathway

A second experimental system we will describe is the imaging of biosynthetic transport and associated regulatory machinery in the secretory pathway. Compared with the visualization of Golgi behaviour during mitosis, these events occur much more quickly. For example, the movement of transport clusters of newly synthesized material from the endoplasmic reticulum to the Golgi complex (14) or from the *trans*-Golgi network (TGN) to the plasma membrane (*Figure 2A* and *B*) occurs at peak velocities of 1–4 μm/sec.

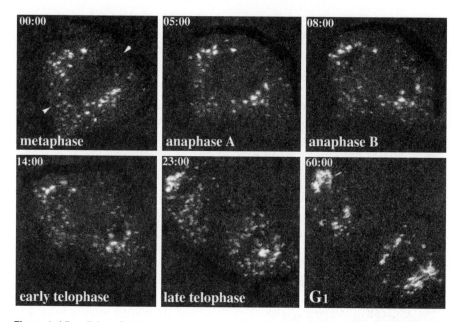

Figure 1. Visualizing Golgi behaviour during mitosis. HeLa cells stably expressing the GFP tagged Golgi marker NAGTI were synchronized at the G_1/S phase of the cell cycle using aphidicolin as described by Shima *et al.* (12). Six hours after release of the cell-cycle block, cells were transferred to carbonate-free medium buffered with 30 mM Hepes and observed by fluorescence microscopy (for more technical details, see text). Images were acquired every 30 sec for 90 min. Several phases of mitosis are shown.

Therefore, in order to track individual membrane carriers and to determine their transport kinetics quantitatively, it is of paramount importance to use equipment that allows the acquisition of images in sub-second intervals. Such high demands on the speed of image acquisition become even more important when more than one fluorescent marker has to be followed. Since the structures under observation move so quickly, sequential imaging of the differently labelled markers may cause a distortion in their spatial relationship, since they may move significantly during image acquisition. Simultaneous image acquisition at different excitation wavelengths, as is possible with a confocal microscope, helps to overcome this problem. However, the rate of image acquisition with a confocal microscope is also limited (approximately 0.5 sec); fast-moving structures are difficult to identify, and transport kinetics are difficult to determine. In summary it is therefore fair to say that the technical possibilities for acquiring multicolour images at a speed required to analyse all aspects of membrane transport in the secretory pathway are still limited.

In order to overcome some of these limitations when studying transport kinetics in the secretory pathway, we therefore use a set-up with a cooled CCD camera (see above), as this microscope system allows image acquisition

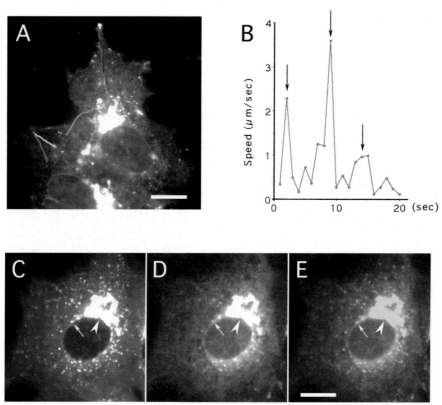

Figure 2. Visualizing and quantifying the kinetics of GFP-tagged vesicular stomatitis virus ts-O45 glycoprotein (ts-O45-G) and COPI-coated transport intermediates in the secretory pathway. (A and B) Plasmid DNA encoding ts-O45-G was injected into the nucleus of Vero cells (see *Protocol 4*). Subsequently cells were incubated at 39.5°C for 4 h to accumulate newly synthesized ts-O45-G in the endoplasmic reticulum (ER), then cells were transferred into carbonate-free medium buffered with 30 mM Hepes and further incubated for 2.5 h at 20°C to accumulate ts-O45-G in the *trans*-Golgi network. The temperature was then increased to 31°C to release the transport block and ts-O45-G transport to the plasma membrane was imaged by acquiring images every 0.5 sec for 5 min. The blue illumination light from the microscope's light source was attenuated 50-fold in order not to interfere with transport. (A) Two cells at the beginning of the observation period. The overlays mark the trajectories taken by distinct ts-O45-G transport intermediates *en route* to the plasma membrane. (B) Quantitative evaluation of the movements of the transport intermediate marked by the red overlay in (A). The transport carriers typically move to the plasma membrane in steps, which are characterized by peak velocities (marked by arrows) between 1 and 4 μm/sec. (C and D) Cells in which GFP-tagged ts-O45-G had accumulated in the ER at 39.5°C (see A and B) were micro-injected at this temperature with Cy3 labelled anti-COPI antibodies (see section 3.2) in order to fluorescently label COPI. Subsequently, the temperature was shifted to 31°C and the transport kinetics of ts-O45-G and COPI-coated structures were imaged in parallel by sequential acquisition of images at the two respective wavelengths for 2 min. (C) Fluorescently labelled COPI; (D) the corresponding image for ts-O45-G and (E) merged image of (C) and (D). Ts-O45-G transport clusters are COPI coated in transit from the ER to the Golgi complex. All images shown were acquired with the microscope set-up using a cooled CCD camera (see above). Bars represent 20 μm.

at a frequency of up to four images per second. To quantify the movements of transport intermediates we then use an extension to the IPlab software, developed in our laboratory, which allows interactive tracking, quantification and visualization of trajectories of individual transport intermediates (see *Figure 2*; copies of this software extension may be obtained on request). In experiments where the precise spatial relationship of two or more fluor-escently labelled markers needs to be determined, we routinely use a confocal microscope.

Figure 2A and *B* show an example of TGN to plasma membrane transport of GFP-tagged vesicular stomatitis virus ts-O45 glycoprotein. *Figure 2C–E* show a double-labelling experiment to determine the transport kinetics and relationship of membrane transport intermediates and the vesicular coat complex, COPI.

References

1. Hermanson, G.T. (1996). *Bioconjugate Techniques.* Academic Press, London.
2. Pepperkok, R., Saffrich, R. and Ansorge, W. (1998). In *Cell Biology:A Laboratory Handbook* (ed. J. Celis), p. 23. Academic Press, London.
3. Chalfie, M., Tu, Y., Euskirchen, G., Ward, W.W. and Prasher, D.C. (1994). *Science*, **263**, 802.
4. Heim, R., Cubitt, A.B. and Tsien, R.Y. (1995). *Nature*, **373**, 663.
5. Ellenberg, J., Lippincott-Schwartz, J. and Presley, J.F. (1998). *Biotechniques*, **25**, 838.
6. Zernicka-Goetz, M., Pines, J., Hunter, S.M., Dixon, J.P.C., Siemering, K.R., Haseloff, J. and Evans, M. J. (1997). *Development*, **124**, 1133.
7. Cole, N.B., Smith, C.L., Sciaky, N., Terasaki, M., Edidin, M. and Lippincott-Schwartz, J. (1996). *Science*, **273**, 797.
8. Llopis, J., McCaffery, J.M., Miyawaki, A., Farquhar, M.G. and Tsien, R.Y. (1998). *Proc. Natl Acad. Sci. USA*, **95**, 6803.
9. Miesenbock, G., De Angelis, D.A. and Rothman, J.E. (1998). *Nature*, **394**, 192.
10. Pepperkok, R., Squire, A., Geley, S. and Bastiaens, P.I.H. (1999). *Curr. Biol.*, **9**, 269.
11. Sullivan, K.F. and Kay, S.A. (eds) (1999). In *Methods in Cell Biology*, Vol. 58: *Green Fluorescent Proteins*, p. 1. Academic Press, San Diego, CA.
12. Shima, D.T., Haldar, K., Pepperkok, R., Watson, R. and Warren, G. (1997). *J. Cell Biol.*, **137**, 1211.
13. Celis, J. (ed.) (1998). *Cell Biology:A Laboratory Handbook*. Academic Press, London.
14. Scales, S.J., Pepperkok, R. and Kreis, T.E. (1997). *Cell*, **90**, 1137.
15. Shima, D.T., Cabrera Poch, N., Pepperkok, R. and Warren, G. (1998). *J. Cell Biol.*, **141**, 955.
16. Yang, W., Pepperkok, R., Bemder, P., Kreis, T.E. and Storrie, B. (1996). *Eur. J. Cell Biol.*, **71**, 53.

6

Visualizing fluorescence in
Drosophila—optimal detection in
thick specimens

ILAN DAVIS

1. Introduction

Drosophila has been a key model organism for studying the genetic control of development. Its success is largely due to its powerful genetics combined with the ability to visualize specific gene products in a beautiful multicellular cytology. This often involves detecting proteins and mRNA deep in whole-mount specimens such as embryos, egg chambers or imaginal discs. Such experiments can be technically challenging, since most objective lenses are designed to image sources of light located just below the coverslip. Imaging fluorescence at some distance from the coverslip creates a number of distortions which reduce the resolution and lower the signal-to-noise ratio.

The use of green fluorescent protein (GFP) has revolutionized our ability to follow the distribution of proteins in living specimens. However, imaging fluorescence in living cells is even more challenging than in fixed material since living specimens must be mounted in media that are compatible with cell viability. Such media have relatively low refractive indexes (RIs) which worsen the imaging difficulties in thick specimens.

In this chapter, I describe simple methods used in my lab to prepare fixed or living *Drosophila* embryos and egg chambers for optimal imaging of fluorescent signals. I explain how to identify and correct aberrations created in thick specimens by imaging fluorescent beads. Many of the protocols described in this chapter are based on methods developed by various investigators over several years and their origins are indicated throughout the chapter. However, there is clearly much scope for improving the protocols; I encourage you to experiment. Many of the issues dealt with in this chapter are also important for other model systems and the techniques can be adapted easily for use in other organisms.

2. Spread of light from a point source

A comprehensive description of optical theory and microscope design is beyond the scope of this book and is covered elsewhere (1,2). Here, an intuitive explanation of how light spreads within a specimen and how optical aberrations arise is more appropriate (see also Chapter 4). It is useful to think of a fluorescent specimen as a three-dimensional (3D) array of point light sources, each with a specific intensity separated by 0.1 µm, a distance smaller than the theoretical limit of resolution of the light microscope (see caption to *Figure 1*). Such an array constitutes the real image and is what we ultimately attempt to visualize with the light microscope. However, when an objective lens creates an observed image of one section within the 3D array of light sources, it also captures out-of-focus light from the neighbouring sections and can introduce a number of optical aberrations or distortions (*Figure 2b–e*).

An important simplifying concept is that the spread of out-of-focus light from each point in the specimen can be considered independently. It is therefore sufficient to describe the way in which out-of-focus light spreads from a single point source of light, such as a 0.1 µm fluorescent bead just below the coverslip, in order to describe the spread of light in the whole specimen. This approach is the key to how out-of-focus light can be re-assigned to its correct position by a computational method known as deconvolution (section 4). However, beads are extremely useful for all kinds of other fluorescent imaging applications including the use of a laser-scanning confocal microscope (confocal microscope). A second important concept which is the basis of many of the protocols described in this chapter, is that every effort should be made to reduce the presence of optical aberrations. This results in much less out-of-focus light in the observed image and greatly simplifies the task of determining the real image by deconvolution or with a confocal microscope.

2.1 Basic properties of objective lenses

Light travels at different velocities in different media and is refracted (changes direction) when passing from one medium to another. The refractive index (RI) of a medium is the ratio of the velocity of light in vacuum (in practice, air) to its slower velocity in that medium. The change in direction is directly related to the RIs of the two media and is governed by Snell's law (see caption to *Figure 1b*). Lenses exploit this diffraction to focus a point light source to form an image (*Figure 1a*). The RI varies with temperature and wavelength, so objective lenses are made of many elements of different kinds of glass to compensate for these variations and other aberrations (*Figure 2b–e*).

The two most important parameters describing objective lenses are the magnification and the numerical aperture (NA). The NA is a measure of the light-gathering capacity of the lens which is dependent on the cone angle (α) and refractive index (η) of the medium between the lens and the coverslip (see *Figure 1a* and caption, and *Figure 2a*). The NA of a lens determines the brightness of the image and the resolving power in both the vertical (z-)axis and the x–y plane of focus (see caption to *Figure 1*). The NA of a lens is always smaller than the RI of the immersion medium. In general, lenses with higher NAs use media with higher RIs.

2.1.1 Classes of objective lens

There are four main classes of high-performance lenses, suitable for fluorescence microscopy:

(a) dry lenses, where air is present between the objective and coverslip. These are generally lower magnification lenses with a low NA, such as 20x/NA 0.75 or 40x/NA 0.85.

(b) oil-immersion objectives in which immersion oil is placed between the objective and coverslip. The RIs of the oil and the coverslip glass are very similar, so the coverslip can be thought of as the last optical element of the oil-immersion objective lens. Such lenses are designed to image objects which are just below the coverslip and are the highest performing lenses, with the highest available NA. There is an absolute limit to the highest available NA of about 1.4 for 60× or 100× objective lenses since a higher cone angle (α) would give rise to reflection of the light at the coverslip.

(c) water-immersion lenses; some of these are coverslip corrected, while others do not use a coverslip. Such objectives cannot achieve the same high NA as oil-immersion objectives. However, they are much less susceptible to aberrations when imaging water-mounted samples at a distance from the coverslip.

(d) glycerol-immersion lenses, which are used with glycerol between the coverslip and objective. Under some circumstances these lenses can be used without a coverslip.

2.1.2 Resolution and brightness of objective lenses

Resolution is defined as the distance between two point light sources at which they can just be distinguished as two separate points rather than a single point. The Rayleigh criterion for two points to be resolved is for the intensity between the two points to be ≤80% than the intensity at the centre of each point (see also Chapter 4). The theoretical x–y resolution (in the plane of focus) is given by a simple formula which depends on the wavelength and NA of the lens: d, the smallest resolvable distance, is equal to 0.61λ/NA. In practice, this will vary considerably depending on the degree of coherence of the light, the resolution of the detector and how many aberrations are present. The z (axial) resolution also depends on the NA and is always lower than the x–y resolution. In the case of a 1.4 NA lens, the best theoretical z resolution can only be 1.7 times poorer than the x–y resolution. Lower NA lenses have an even poorer ratio of z to $x–y$ resolution, which depends on the NA. Spherical aberration reduces z resolution considerably, but the mounting method can be equally important to the quality and resolution of the imaging (*Figures 3* and *6*).

Measuring the actual resolution achieved by a given lens and microscope system in a particular imaging situation is not an easy task if the Rayleigh criterion for resolution is used. Two identical point light sources are difficult to place at desired distances apart. An alternative criterion for resolution which can be applied to a single point light source is the apparent full-width–half-maximum (FWHM) size of a fluorescent bead, i.e. the width at which the intensity of fluorescence from a point source of light is at 50% or more of the maximum intensity at the centre of the bead. The FWHM is a very simple parameter to measure from images of a 0.1 µm bead and can be used to compare the performance of a given lens in different imaging circumstances in $x–y$ and z (see *Protocol 3* and *Figure 4*).

The intensity of a fluorescent image captured from a point source of light depends on many factors. These include the absorption and reflection qualities of all the optical elements in the light path and the wavelength of light. Assuming all these factors are equal then the intensity depends primarily on the NA and magnification of the objective lens. The intensity I is proportional to the fourth power of the NA and inversely proportional to the magnification (mag) squared:

$$I \alpha\ NA^4/mag^2$$

In practice this means that the brightest possible image is obtained with a 60× or 63× lens with an NA of 1.4. In addition, the relative brightness of two different lenses can be calculated, assuming all other factors are equal. For example a 100x/NA 1.4 lens will be 3.8 times brighter than a 100x/NA 1.0 objective.

2.1.3 Optical aberrations and their correction

All objective lenses are subject to a variety of optical aberrations which are more significant in the case of lenses with high NA and high magnification. Lens design and manufacturing techniques have improved considerably, so that today's best high NA lenses are corrected to a very high degree for most aberrations when used under optimal conditions. However, objective lenses corrected for all possible aberrations are not available and imaging fluorescent signal in thick specimens creates a number of challenging aberrations. There are three main types of aberrations which are important to consider (*Figure 2b–e*).

The first and most important aberration is spherical aberration, which results in light rays at different distances from the centre of the lens focusing to different points (*Figure 2b*). Imaging light sources in thick specimens creates substantial spherical aberration, especially with oil-immersion objectives. For

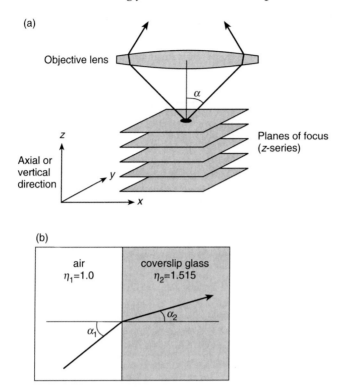

Figure 1. An explanation of how objective lenses work. (a) An objective lens gathers light from a point light source at a particular focal plane (*x–y* plane) at a fixed distance from the lens. A 3D image of a specimen consists of a series of focal planes at varying position along the *z*-axis. The objective lens focuses the light by refraction (see panel b and *Figure 2a*) to form an image within the microscope. The cone angle (α) determines how much light can be gathered from the light source and is related to the NA of the lens. (b) Light is refracted (changes direction) when it passes from air into glass, since its velocity is different in different media. The refractive index of a material is the ratio of the velocity of light in that medium over its velocity in a vacuum (or air). Snell's law describes the relationship between the refractive index and angle describing the direction of travel of the light waves within the medium: $\eta_1 \sin\alpha_1 = \eta_2 \sin\alpha_2$. The NA of a lens is related to α and the refractive index η of the medium between the objective lens and the coverslip: $NA = \eta \sin\alpha$.

this reason this chapter deals with practical methods of reducing spherical aberration, in order to improve the quality of fluorescent imaging and differential interference contrast (DIC) imaging (3) . Some lower NA objectives have an adjustable collar which compensates for coverslip thickness and can be used to correct spherical aberration. The higher NA lenses do not have such collars, so the only practical alternative is to use an immersion oil with a different RI (see section 2.5).

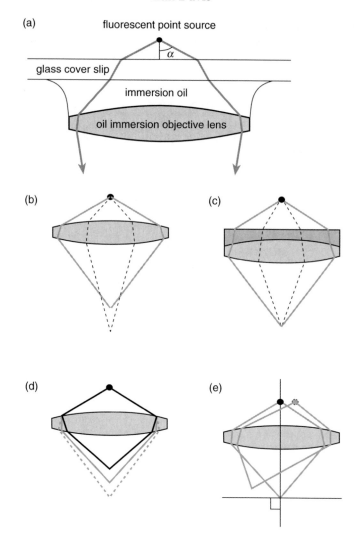

Figure 2. (a) The path of light in an oil-immersion objective used with an inverted microscope. Light from a fluorescent point source first passes through the medium in which the specimen is mounted and is refracted when passing through the coverslip, immersion oil and the front element of the lens. The oil-immersion objective performs optimally when imaging an object just below the coverslip surface. (b–e) Major aberrations present in objective lenses. (b) Spherical aberration causes light rays from the same point which pass at different distances from the centre of the lens to be focused at varying distances from the objective. (c) Spherical aberration is corrected in objective lenses by including many lens elements made of different kinds of glass with different refractive indices. (d) Chromatic aberration causes light of different wavelengths to be focused at varying distances from the objective and is also corrected by using multiple lens elements. (e) Field curvature is noticeable in non-plan lenses, causing light from different positions in the same focal plane to be focused to different parts of a curved surface rather than a flat image plane.

The second aberration is chromatic aberration, which is caused by light of different wavelengths being focused to different points by the lens (*Figure 2d*). Most lenses are corrected for chromatic aberration in blue, green and red light. However, there is often very small residual chromatic aberration and chromatic aberration is often quite pronounced in UV and far-red, in both z and x–y. Chromatic aberration can be accurately measured by using multi-wavelength fluorescent beads and can be corrected after the images are captured by shifting the relative registration of the images of different wavelengths.

The third important aberration is field curvature (*Figure 2e*). This aberration is very pronounced in non-plan lenses. However, in plan lenses field curvature is very small and can be ignored when imaging near the centre of the field.

2.2 Measuring the spread of light within a specimen

Some specimens contain only a few very bright points of fluorescence which approximate point light sources so that the spread of light in the specimen can be imaged directly. However, in the majority of cases fluorescent images result from a complex 3D array of light sources of differing intensities. Ideally, bright fluorescent beads would be injected into a control non-fluorescent specimen to visualize the spread of light from a single point source within the specimen. A much more practical alternative is the use of fluorescent beads embedded in a similar medium to the specimen. Beads labelled with multiple fluorochromes are particularly useful if multiple wavelengths are co-visualized.

Protocol 1.   Making fluorescent bead slides

Equipment and reagents

- Glass slides
- 22 mm × 22 mm no. 1.5 coverslips (approximately 0.17 mm thick)
- 50 °C hot-plate or oven
- 0.1 and 0.5 μm multi-wavelength beads (TetraSpeck; Molecular Probes T-7284)

- Red intensity calibration beads (InSpeck kit; Molecular Probes I-7224)
- Absolute ethanol
- Mountant
- Nail varnish

Method

1. Dilute the mixture of different size beads desired (e.g. 0.1 and 0.5 μm Tetrabeads) 500-fold[a] into pure ethanol.

2. Pipette 1 μl of diluted beads on to a coverslip and on to the centre of a glass slide, and spread over the entire surface with a yellow pipette tip. Dry on a 50 °C hot-plate or oven for 20 min.

3. Mount the coverslip (bead side down!) on to the glass slide (bead side

---

Protocol 1. *Continued*

up!) using an appropriate mountant which is used in your experiment.[b] Choose an appropriate volume of mountant which will result in the distance between the coverslip beads and the slide beads being approximately the same as the depth of the fluorescent signal imaged in your real sample.

4. Seal the coverslip with nail varnish to prevent evaporation and movement of the coverslip over time. Keep the slides in the dark at 4°C (do not freeze).

---

[a] This dilution will vary between batches and manufacturers, and will have to be determined empirically.
[b] See *Table 3* for details of mounting media.

2.3 Determining the point-spread function of a particular lens and microscope configuration

The point-spread function (PSF) is the mathematical equation that describes how out-of-focus light spreads from a point source of light. It can be measured by imaging a single 0.1 μm fluorescent bead in 3D space at different wavelengths. The wave nature of light results in so called 'Airy rings' of out-of-focus light above and below the bead (*Figures 3* and *5*). In an ideal microscope the light spreads exactly symmetrically above and below the bead with the size and number of concentric rings increasing and the intensity decreasing sharply with distance from the bead. However, each individual objective lens has a slightly different PSF, even from the same batch of equivalent lenses. The bead slide can be used to determine the PSF of the particular lens and microscope configuration, and poorly manufactured or damaged lenses can be identified.

The PSF captured from a single 0.1 μm fluorescent bead contains considerable information which is best captured and analysed with a digital microscope system such as a CCD camera or confocal microscope. However, if you only have a fluorescent microscope with conventional photography, you may still be able to obtain useful information from viewing beads by eye or photographing them. In such circumstances it is probably best to use the much brighter 0.5 μm beads by focusing up and down and judging, by eye, the way the light spreads from the bead, to identify spherical aberration. A calibrated eyepiece graticule can be used to measure distances in the x–y plane. It should be noted that some confocals may not be sensitive enough to capture useful images from the smallest beads and may necessitate the use of the larger beads or non-optimal imaging conditions such as a wide open pinhole and very high detector amplification, which introduces noise.

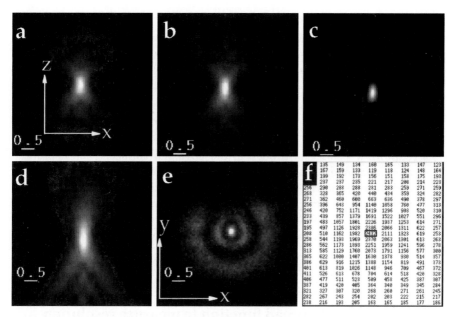

Figure 3. Examples of PSF of 0.1 μm fluorescent beads mounted in glycerol and imaged in the rhodamine channel. Panels a–d are shown in side view (*x–z*). (a) PSF of a bead mounted just below the surface of the coverslip and imaged with 1.515 immersion oil, showing no visible spherical aberration and some out-of-focus light. (b) A similar bead imaged at a depth of 15 μm below the coverslip, using 1.524 immersion oil. There is only very slight spherical aberration and slightly more out-of-focus information than in (a). (c) The results of deconvolving the image shown in (b) using a PSF of a bead at the surface of the coverslip, showing very little out-of-focus information. (d) A similar bead to that in (b) imaged at a depth of 15 μm below the coverslip, using inappropriate immersion oil (1.514). Note the increase in the size of the bead in the *z*-direction and the large Airy rings below the bead, caused by spherical aberration. (e) An *x–y* view of out-of-focus Airy rings showing a missing segment, indicating that a bubble is present in the immersion oil. (f) The actual greyscale values representing the intensity of light at each pixel around the centre of the bead shown in (b). These values have been plotted in a graph shown in *Figure 4*. The scale bar represents 0.5 μm.

Protocol 2. Determining a point-spread function using fluorescent beads

Equipment and reagents
- bead slide (see *Protocol 1*)

1. View the slide using a high NA lens with the standard immersion oil supplied by the manufacturer. This will be optimized for imaging just below the level of the coverslip. Find a bright bead (0.5 μm) just below the coverslip and make a note of the *z* position.

2. Shift the focus (*z* position) until similar bright beads are in focus on the

Protocol 2. *Continued*

surface of the slide. Make a note of the *z* position and calculate the total depth from the surface of the coverslip to the surface of the glass slide.

3. Focus back to a bead on the coverslip surface and find a small (0.1 μm)[a] bead which is as bright as possible but is a point light source and does not have a discernible structure. Make sure the bead is well separated from other beads so that out-of-focus light from neighbouring beads is not visible.

4. Capture a *z*-series of a bead at the coverslip surface. Then capture another *z*-series of a bead at a depth which you would use to capture data from your 'real' sample. Attempt to maximize the resolution by turning pixel binning off in the CCD camera and use your highest NA lens such as 100x/NA 1.4. In the case of a confocal microscope, capture a *z*-series under a number of conditions such as pinhole size and PMT settings.

[a] If the imaging system is not sensitive enough, then 0.5 μm beads should be used instead (see above).

2.4 Using a point-spread function to measure resolution and optical aberrations

Once you have captured a PSF, the stack of images contains considerable information which is best analysed using a digital image analysis system. If you lack such a system, some simple suggestions of how to utilize the information from a bead are given in the previous section.

Protocol 3. Measuring resolution, spherical and chromatic aberrations

Equipment and reagents
• bead slide (see *Protocol 1*)

1. Capture a PSF from a 0.1 μm bead as described in *Protocol 2* for the wavelengths you are likely to be imaging. Use image analysis software to view the intensity of light (in terms of greyscale values) at different focal planes. View the Airy rings above and below the centre of the bead. If the rings appear as if a segment has been removed from one side, then a bubble was probably present in the immersion oil or close to the sample and a new PSF should be captured (see *Figure 3e*).

2. Find the exact focal plane of the bead, where the intensity is maximal. Note that even a 0.1 μm change in *z* will affect the intensity.

3. Determine the diameter of the circle within which the intensity is ≥50% of the maximum intensity in the centre of the bead (FWHM).[a] This represents the apparent size of the bead and is a reasonable measure of the *x–y* resolution (see *Figure 4*).

4. Move to planes of focus above and below the focal plane of the bead and determine the distance in z within which the intensity of the light remains $\geqslant 50\%$ of the maximum intensity in the centre of the bead (FWHM). This distance represents the apparent size of the bead in z and is a reasonable measure of the z resolution. In the best possible cases using an NA 1.4 lens, the z resolution will never be less than 1.7 times the $x-y$ resolution. Usually this number is greater, especially if a lower NA lens is used or if there is a lot of spherical aberration (see *Figure 4*).

5. Rotate the stack of images by 90° in the x-axis. This will allow you to view the PSF of a bead from the side. If this facility is not available you can achieve the same results by moving above and below the plane of focus of the bead and noting the appearance of the Airy rings. If the spread of light is symmetrical above and below the bead, then spherical aberration is small. However, if stronger Airy rings are visible above than below the bead plane of focus (or vice versa) then the imaging suffers from spherical aberration. Spherical aberration has a profound effect on z resolution and a smaller effect on $x-y$ resolution.[b]

6. Compare the appearance of the bead in different wavelengths to determine the extent of chromatic aberration. This is easily seen if the different wavelengths are co-visualized as a three-colour image. Chromatic aberration can result in a significant shift in $x-y$ or in z between different wavelengths imaged in multiply labelled samples. It is particularly apparent in the case of poorly corrected objectives (see section 5). However, even highly corrected lenses tend to have a lack of registration of at least 0.2 μm between different wavelengths.

7. Capture a PSF for a number of beads in a large field. Examine a bead in the centre of the field and compare it with one at the edge of the field. Determine if the focal plane at which the maximal intensity is found, is the same for the two beads. In addition, determine whether the Airy rings are symmetric about the centre of the bead at the edges. This is a measure of the curvature of the field and is a property of the lens used.[c]

8. Use a bead slide consisting of a mixture of beads with known different intensities (Molecular Probes) to measure the ability of a microscope to discern objects of different intensities.[d]

[a] The best results are achieved by plotting a curve of intensity against distance or number of pixels (see *Figure 4*).
[b] For example, imaging fluorescence deep in a specimen can decrease the z resolution from 0.5 μm to 2 μm, when using a 100x/NA 1.4 lens.
[c] This kind of aberration is negligible in plan lenses which, as their name suggests, are designed to have a very flat field.
[d] This will depend on many factors including the sensitivity of the microscope and the dynamic range of the detector.

143

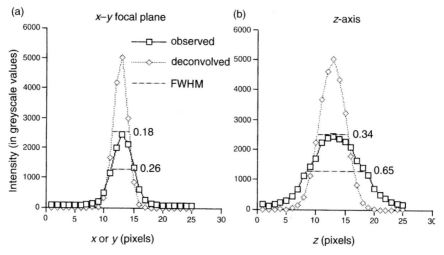

Figure 4. Graphs of the greyscale values of pixels along a line through the brightest part of the beads shown in *Figure 3* panels b (observed) and c (deconvolved). (a) Plots of values along a line parallel to the *x*-axis. (b) Plots of values along a line parallel to the *z*-axis. The full-width–half-maximum value (FWHM) is shown in grey figures beside each curve. The curves show that *x–y* FWHM is reduced from 0.26 μm before deconvolution to 0.18 μm after deconvolution. *z* FWHM is reduced even more dramatically, from 0.65 μm to 0.34 μm, by deconvolution. Therefore, deconvolution improves the *x-y/z* resolution ratio from 2.5 to 1.9.

2.5 Minimizing spherical aberration by varying the refractive index of the immersion oil

Spherical aberration is an important factor reducing the quality of images acquired in thick specimens (see previous sections). Some lower NA lenses have coverslip thickness correction brackets which can be used to correct spherical aberration, but high NA lenses tend to lack this feature: a number of water-immersion lenses have been introduced recently, however, which do have these properties, but which are expensive. Often, the only practical alternative is to choose an appropriate immersion oil with an increased RI to correct the spherical aberration.

Table 1. Examples of optimal immersion oil to minimise spherical aberration

Mounting medium	(RI)	Depth	Optimal immersion oil (RI)
glycerol	(1.466)	0 μm	1.516
glycerol	(1.466)	5 μm	1.518
glycerol	(1.466)	15 μm	1.524
halo-carbon oil series 95	(1.408)	15 μm	1.534
vectashield	(1.466)	20 μm	1.534

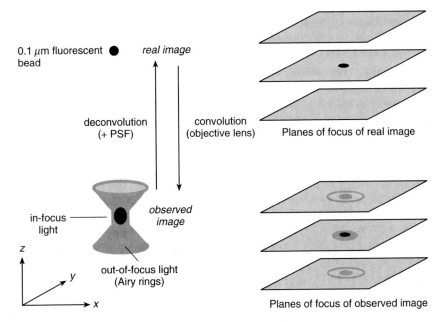

Figure 5. (a) Diagram illustrating the principles of convolution and deconvolution. The real image is a stack of *x–y* images (*z*-series) which represents the actual intensity of light present in a specimen. In this example, the real image of a 0.1 μm fluorescent bead is represented by a single point of signal in one focal plane only. The microscope produces an observed image (also a *z*-series) that results from a process of convolution, which creates out-of-focus light around the bead and in neighbouring *z*-sections (focal planes). The convolution process distorts the appearance of the bead into a series of Airy rings which appear like an hourglass. Deconvolution is a computational process by which the real image is estimated based on the observed image and a PSF of a 0.1 μm bead captured at the surface of the coverslip. It is an iterative process which results in the re-assignment of out-of-focus information in the observed image to its point of origin.

Protocol 4. Optimizing the immersion oil to reduce spherical aberration

Equipment

- Immersion oils with different RIs (Cargille Labs Inc.), or Cargille oil kit (RIs 1.512–1.534) sold by Applied Precision Inc.
- 0.1 and 0.5 μm fluorescent-bead slide (*Protocol 1*)

Method

1. Capture a PSF of a small bead (0.1 μm) mounted in the medium used to mount the embryos or egg chambers and at the depth at which the fluorescent signal will be imaged. Start with the standard immersion oil recommended by the manufacturer, for example R.I. of 5.18 for Zeiss or 5.16 for Olympus. Rotate the stack of images by 90° in the *x*-axis.

Protocol 4. *Continued*

With experience, you can simply inspect the spread of light from a brighter 0.5 μm bead by direct visual observation.

2. Determine which way the spherical aberration is skewing the spread of light from the bead. The direction of the skew indicates whether you should increase or decrease the RI of the immersion oil in order to reduce spherical aberration. If the Airy rings spread away from the lens, then increase the RI of the oil, and vice versa (*Figure 3d*).

3. Try another immersion oil, with a very different RI (e.g. 1.534). This should skew the Airy rings in the opposite direction (e.g. below instead of above the bead).

4. Keep trying oils of an intermediate RI until you find an oil which gives a symmetrical PSF. Notice that overall this oil will also give the least intense Airy rings. In other words this will result in much less out-of-focus light above and below the bead. In addition, The *z* resolution should be markedly increased and the apparent intensity of the bead higher, leading to a better signal-to-noise ratio.

3. Imaging fluorescent signals in fixed *Drosophila* embryos and egg chambers

So far, this chapter has dealt with how to prepare and image fluorescent beads under idealized conditions to reduce potential aberrations created in real specimens. There are a number of additional imaging problems which can arise with thick specimens. This section describes these difficulties and methods for preparing and mounting specimens which will minimize them.

3.1 Difficulties imaging fluorescent signal inside a biological specimen

All biological specimens have significant autofluorescence which varies with wavelength and methods of fixation. In most specimens, increasing the wavelength reduces the autofluorescence considerably. For example, in *Drosophila* embryos yolk autofluoresces much more strongly in the blue or green channel than in the red channel. However, in many experiments the background fluorescence due to the labelling procedure itself is more intense than autofluorescence. Such background can be reduced by optimizing the fixation conditions or by using different antibody combinations. For example, strong glutaraldehyde fixation leads to very bright background fluorescence in the green and red channels. The critical parameter is the signal-to-noise ratio, so increasing the signal by altering the procedure may improve the image. The signal may also suffer from 'photobleaching with long exposures or many *z*-sections. Photobleaching can be reduced by using anti-fade reagents.

Biological specimens are not entirely transparent even after fixation and mounting in a glycerol-based medium. This problem increases with lower wavelengths but is largely dependent on the nature of the particular specimen. Specimens can absorb, reflect or refract light (1). These effects are usually minor in *Drosophila* and can be minimized by reducing spherical aberration and using appropriate mounting and imaging methods (*Protocols 4, 5, 6* and *8*).

3.2 Mounting fixed *Drosophila* embryos and egg chambers

Drosophila embryos have a cylindrical diameter of approximately 100–150 μm, which varies depending on how the embryos are processed and mounted. If a glancing section at the surface of the embryo is required, then the signal will emanate from a depth of approximately 10 μm below the surface. In this case, the embryos should be mounted so that the coverslip is touching the specimen. However, many experiments require the acquisition of images in cross-section through the middle of the embryo. In this case fixed embryos should be mounted so that they are considerably squashed to minimize the thickness of tissue and mounting medium to approximately 30–40 μm (see *Protocol 5*). Most stages of egg chambers are considerably smaller than embryos, so fixed egg chambers present fewer imaging problems than embryos. However, in some cases it is also useful to mount egg chambers under conditions which flatten the tissue.

Protocol 5. Mounting fixed *Drosophila* embryos and egg chambers

Equipment and reagents

- Etched tungsten needles (Interfocus Ltd)
- PBT: phosphate-buffered saline (PBS) with 0.1% Tween 20
- P1000 pipettor
- Blue and yellow pipette tips
- 22 × 22 mm coverslips
- Mountant
- Nail varnish

Method

1. After completing the staining protocol (see *Table 2* for different protocols) wash the embryos or egg chambers in PBT to remove excess stains or antibodies.

2. Take up approximately 10–20 μl of embryos into a blue pipette tip in as small a volume of PBT as possible. Pipette the embryos on to a glass slide.

3. Remove as much of the PBT as possible by placing a P1000 with a blue tip in the centre of the drop of embryos (exactly perpendicular to the glass slide) and sucking the excess liquid from several different positions. Do not let the embryos dry out.

Protocol 5. *Continued*

4. Pipette approximately 30 μl of mounting medium with a cut-off yellow tip on to the embryos. Gently mix very thoroughly with a fresh intact yellow tip, taking care to avoid creating bubbles. Ensure that there is only a single layer of embryos. Note that bubbles can be pipetted away with a yellow tip and that the embryos do not easily break if fixed sufficiently.

5. Gently place a clean 22 × 22 mm coverslip on the embryos, taking care to avoid creating bubbles. The drop of liquid and embryos should spread slowly underneath the coverslip, with no liquid to spare.[a]

6. Seal the coverslip with nail varnish to prevent evaporation and movement of the coverslip over time. Keep the slides in the dark at 4°C (do not freeze).

[a] The exact amount of liquid added depends on the number of embryos and has to be determined by the investigator. I find that 20–30 μl works well.

Table 2. Protocols for immunostaining fixed *Drosophila* specimens

Protocol	Reference
Immunolabelling embryos	
Whole mount embryos	(4)
Flat dissection of embryos	(4)
Fluorescent mRNA in situs	(5)
Cytoskeleton	(6)
Nervous system	(7)
Rhodamine-Phalloidin staining	(8)
Immunolabelling egg chambers	
Antibody labelling	(4)
Rhodamine-Phalloidin staining	(9)
β-galactosidase staining	(9)
Immunolabelling larvae	(7)
Imaginal discs	(4)
Immunolabelling of salivary glands	(10)
Immunolabelling of sections	(4)
Double and multiple labelling	(4, 7)
Labelling problems	(4)

3.3 Choosing the correct mounting medium

In general, oil-immersion objective lenses are best used with specimens mounted in media with the highest possible RI (*Table 3*), although other considerations may also be important. These include photobleaching (80% glycerol, without antibleaching reagents), the presence of some background fluorescence in the UV channel (Vectashield, Vector Labs), slow evaporation

Table 3. Refractive indices of common mounting media

Media	(RI at 23 °C)
Air	1.000
water	1.350
coverslip glass	1.515 varies with wavelength and temperature.
other kinds of glass	dependent on the composition of the glass, wavelength and temperature.
Mounting media	
Halo-carbon oils	
series 27	1.4059
series 95	1.4082
series 700	1.4120
Glycerol:	
100% glycerol	1.466
80% glycerol	1.446
50% glycerol	1.406
40% glycerol	1.391
Immersion oil:	
Zeiss	1.518
Olympus	1.516
Fluoromount-G	1.3929
FluorGuard	1.470
DABCO	1.470
PS-Speck (Molecular Probes)	1.470
Mowiol	1.490
Vectashield	1.440
API oil kit (Cargille oils)	1.512–1.534 (in 0.002 increments)

of mountant even when painted with nail varnish and kept at 4°C
(Fluoromount) or inhomogeneity and lack of reproducibility (Mowiol).

4. Re-assigning or removing out-of-focus light

Out-of-focus light is an inevitable consequence of the design of conventional
fluorescence microscopes. The specimen is illuminated in more than just the
focal plane of interest, so that the resulting image represents the summation
of all the Airy rings of points above and below the focal plane, as well as the
sharp image from the focal plane itself. There are currently three main
methods of reducing the effect of out-of-focus light (11).

The first, which is dealt with in this chapter, is by computationally re-
assigning the light to its correct point of origin in a 3D stack of images, a
process known as deconvolution. This does not depend on any major modifications to the microscope design, but is the hardest to understand intuitively.
The best results are achieved when fluorescent beads are used to generate a
PSF for each particular high NA lens and to reduce spherical aberration as

described above. In general, this method achieves the highest resolution and is more quantitative than the other two methods described below. However, it must be stressed that using beads to reduce spherical aberration improves the resolution in the case of other methods of removing out-of-focus information.

The second method is the most common and involves the use of a very different microscope design. A confocal microscope has pinholes in the correct optical path to prevent the out-of-focus information from arriving at the detector (Chapter 4). Confocal imaging tends to perform better than deconvolution in a thick sample with a very large amount of background and poor signal-to-noise ratio. In contrast, if the sample contains discrete signal in low background (high signal-to-noise ratio), then deconvolution systems tend to perform better than confocal imaging, even in thick specimens. Chapter 4 and ref. 11 provide a more detailed comparison of confocal microscopes and deconvolution systems.

The third method uses multi-photon microscopy with a very high energy pulsating laser (Chapter 4). The basic principle behind such a microscope is that two or more low energy (long wavelength) photons are used to excite a fluorochrome sequentially instead of a single photon in conventional fluorescence. This phenomenon depends on having a sufficient number of photons which can only occur near the plane of focus, using a very powerful laser. Away from the plane of focus, there are insufficient numbers of photons to excite the fluorochrome by a two-photon effect. While multi-photon microscopy is extremely promising, it is a recent technology and has not been used routinely by many labs to image biological specimens. There are a number of potential problems with the technology which will probably be overcome in the next few years (11).

Finally, confocal and multi-photon images have some residual out-of-focus light that can be re-assigned by deconvolution using methods described in this chapter. Furthermore, deconvolution of confocal or two photon images may yield the highest possible resolution (12,13).

4.1 Deconvolution: basic principles

Convolution is the process by which the real image becomes blurred by the addition of out-of-focus light to produce the observed image. The convolution process that affects an 'ideal microscope' is a well understood phenomenon in physics. However, convolution is slightly different in every microscope and varies between different high NA lenses. For a given microscope and lens, the best description of convolution is contained in the PSF of a 0.1 μm fluorescent bead mounted just under the coverslip (see *Protocol 1*).

Deconvolution is the computational process by which the real image can be estimated from the observed image and the PSF of a 0.1 μm bead. There are several very different methods of deconvolution, but all the commercial packages utilize a similar basic approach. This involves an iterative process of calculations which are performed in Fourier space, which is a description of images as a series of wave functions. Fourier space is used simply in order to make the calculations much more rapid, as similar calculations in (x,y,z) space would take very much longer.

The first step of deconvolution is to convert the PSF and the observed image to Fourier space. The Fourier transform of the PSF file is referred to as an optical transfer function (OTF). The program then guesses what the real image is, and convolves it according to the information contained in the OTF. It then compares the convolved guess with the original image and uses the information to make a better estimate. This new estimate is then convolved and the process repeated many times, until there is very little difference between the convolved estimated image and the observed image. A more detailed description of Fourier space and how deconvolution calculations are performed is beyond the scope of this book and is given elsewhere (2).

The mathematical theory for the deconvolution calculation in Fourier space has been understood for many years (14) and the basic technique has been used by NASA for decades. However, the first practical implementation of deconvolution to fluorescent images of biological specimens was undertaken by Agard and Sedat in the early 1980s (15). This implementation is based on 'constrained iterative deconvolution' (14). More recently, a commercial version of the 'Agard–Sedat microscope' and software running on a Silicon Graphics computer has become available, from Applied Precision Inc. (see *Table 7*). Their Delta Vision widefield microscope is probably the best performing and most expensive deconvolution system currently available. However, there are a number of other manufacturers, who make alternative systems, that involve the use of subtly different full 3D deconvolution algorithms developed independently by other researchers. These include 'maximum likelihood deconvolution' by Huygens (sold by Bitplane) and 'exhaustive photon reassignment' by Fred Fay and Walter Carrington (sold by Scanalytics). For a more detailed comparison of some deconvolution algorithms see http://www.videomicroscopy.com/Product%20reviews%20 categorized/Deconvolution/Deconvolution.htm

4.2 Practical considerations

The most important point to appreciate is that the deconvolution software being used is less important than the quality of the observed image and that of the PSF. The quality of the image will depend on the factors described in this chapter as well as the quality of the microscope, objective lens, the alignment and evenness of the illumination and thermal and mechanical stability of the microscope. The specification of the CCD camera used to capture the images may also be important (see Chapters 4 and 5, and http://www. videomicroscopy.com/Summer1998/CCD%20grades.htm). For each high NA lens, a PSF is usually captured by the manufacturer of the deconvolution system under ideal conditions just below the surface of the coverslip using 0.1 μm beads mounted in glycerol. In some circumstances, the user will have to create some of their own PSF files. For example, when an objective has a modifiable NA, a PSF must be captured for each particular NA setting. In the case of low NA lenses, such as 20x/NA 0.75, measuring a PSF is not necessary and a theoretically calculated PSF performs very well. In such cases, the deconvolution process is also less susceptible to imaging problems such as spherical aberration.

In principle, the different deconvolution algorithms should produce very similar results, given the same set of high-quality images and OTFs. However, some commercial packages do not perform a full 3D deconvolution. They may instead only use a nearest-neighbour algorithm or a 2D deconvolution algorithm which only re-assigns out-of-focus light from the closest neighbouring sections or only within a single focal plane, respectively. Such deconvolution 'short cuts' produce much more rapid results, as far fewer calculations have to be performed. However, the degree of re-assignment of out-of-focus light is only limited in comparison with a full 3D deconvolution.

Most of the commercial packages use Silicon Graphics, PCs or Macs to 'number crunch' the images. The one exception is Scanalytics, which uses dedicated hardware (an array processor) designed for rapid deconvolution calculations. The advantages of such a system are the speed of deconvolution and the ability to perform many more iterations. The drawback is that it is expensive and cannot be used for other tasks, such as image analysis.

4.3 Deconvolution of images from thick specimens

Four steps are involved in this process:

(a) A PSF is captured for a 0.1 μm bead at the surface mounted in glycerol and the PSF is converted to an OTF (see *Protocol 2*), a process probably performed by the manufacturer of the deconvolution system.

(b) The image quality is optimized by reducing spherical aberration, by imaging a 0.1 μm bead at the same depth and medium as the specimen (see *Protocol 4*).

(c) The image is captured using the optimal immersion oil, resulting in a *z*-series.

(d) The image is deconvolved using the standard OTF of a surface bead. Capturing a PSF of a bead at the surface tends to produce better results than a PSF of a bead mounted at a similar depth to the specimen.

4.4 Incomplete deconvolution and deconvolution artefacts

If care is taken to minimize spherical aberration and ensure that the microscope system is well aligned, then few major deconvolution artefacts are produced. In general, if the deconvolution is not working well for some reason, then the problem is that not all the out-of-focus information is re-assigned, so the image remains partially blurred (incomplete deconvolution). However, there are always very small or negligible artefacts, which do not have any bearing on the interpretation of the results. These may only be visible if the image is very aggressively contrasted to reveal, for example, very small differences in brightness in the background. The first common artefact is minor variations in the intensity of the background over the whole speci-men giving a speckled appearance. A second minor artefact is a darker region around the brightest sources of light in the specimen. Neither of these artefacts is usually visible when the image is appropriately contrasted. It should be noted that other kinds of artefact can be observed when confocal images are inappropriately contrasted. For example, confocal images are particularly susceptible to noise. In general, it is best to focus the interpretation of the image on major aspects of the image already visible in the original observed image. If possible, a deconvolution CCD imaging system should be used in parallel with a confocal one, to make use of the advantages of both kinds of system.

5. Time-lapse microscopy of living *Drosophila* embryos and egg chambers expressing GFP markers or micro-injected with fluorescent reagents

Visualizing fluorescent molecules in thick living specimens is more challeng-ing than in fixed material, since the mounting medium has a lower RI. Furthermore, the RI of a living specimen is similar to that of water and cannot be modified. In some types of specimen, water-immersion objectives designed to work without a coverslip can solve many of the potential imaging problems such as spherical aberration. However, living *Drosophila* embryos and egg chambers suffer from hypoxia when mounted in an aqueous medium, so they must be mounted in halocarbon oil. If the temperature of the specimen is important, for example when imaging temperature-sensitive mutations, then a very accurate temperature-controlled chamber can be used. I have found ΔT-dishes (made by Bioptechs) to be the most reliable for temperature control. If you need to image a large number of embryos or egg chambers, then it is very useful to be able to film many specimens at the same time. This depends on the imaging system having a motorized stage that can move in the x–y as well as the z plane.

5.1 Mounting, micro-injecting and imaging living embryos

For real-time imaging, the chorion (egg shell) must be removed and embryos (surrounded by an impermeable waxy vitelline membrane) must be cultured in halocarbon oil to avoid hypoxia, as halocarbon oil is extremely well saturated with oxygen. The oil is immiscible with water, so all traces of water must be removed from the surface of the vitelline membrane, otherwise small droplets of water can act like small lenses. In the case of injected embryos, desiccation is the crucial step and must be more severe to reduce leakage of cytoplasm which can alter the results and cause lethality. However, over-desiccation will lead to flacid embryos with folds in the vitelline membrane, giving rise to distorted images and lethality. The desiccation must also be reproducible and consistent from embryo to embryo.

Protocol 6. Preparation and mounting of living *Drosophila* embryos for micro-injection and real-time imaging

Equipment and reagents

- Household bleach or 14% sodium hypochlorite diluted to 50% in water
- Teflon-coated fine watchmaker's forceps (Interfocus 11626-11)
- 24 mm × 50 mm no. 1.5 coverslip (0.17 mm thick) (Laboratory Supplies (UK) Ltd)
- Series 95 halocarbon oil (Halocarbon Products Corporation)
- Double-sided sticky tape dissolved in heptane
- Inverted fluorescence compound microscope[a]

- One fine and one medium paintbrush
- Fine hypodermic needles, or dissection needles
- Small sieve (standard equipment in *Drosophila* labs)
- Agar plates for collecting eggs (standard equipment in *Drosophila* labs)
- Glass scintillation vials (standard equipment in *Drosophila* labs)
- Desiccation chamber: sealed container with self-indicating silica gel (Sigma S-7025)

Method

1. To make the heptane glue, pack a long strip of double-sided sticky tape or packing tape into a glass scintillation vial, fill with 10–20 ml of heptane and leave on a shaker overnight. Pour the heptane into a fresh scintillation vial. Note that the strength of the glue will depend on how much sticky tape and heptane you use. In general, the thinner the glue, the better the imaging will be, as long as the embryos remain attached to the coverslip (see below).

2. Before starting, prepare some coverslips by pipetting some heptane glue on to the coverslips and allowing to air dry.

3. Collect appropriately timed embryos on collection plates (16). Remove dead flies and wash the embryos off the agar using a paintbrush. Pour the embryos into a small sieve and wash thoroughly with water at room temperature to remove the yeast.

4. Dechorionate for 2–5 min[b] at room temperature in a sieve with bleach. Wash thoroughly with water at room temperature to remove the bleach.

5. Dry the embryos thoroughly by pressing a tissue to the bottom of the sieve.

6. Gently pick some of the embryos up with a fine paintbrush and place on to a dried heptane glue coverslip. Once the embryos have stuck to the glue, do not attempt to reposition them, as they will rupture easily.

7. If mounting GFP-expressing transgenic embryos or non-fluorescent embryos for DIC time-lapse microscopy, then desiccate the embryos in a desiccation chamber containing silica gel for 5 min and cover with a small film of halocarbon oil.

8. If the embryos will be micro-injected with DNA or proteins or other reagents, then align on to a small block of apple juice agar (20 mm × 20 mm) placed on a glass slide using a dissecting stereo-microscope. Do not spend too long aligning the embryos, as they will eventually dry out or develop too far. Gently pick up the embryos on to a cover slip coated with heptane glue. Desiccate for approximately 10 min. Determine the optimal desiccation time by trial and error, depending on the exact conditions and method of alignment.

[a] If an upright microscope must be used instead, then alternative slightly less convenient mounting methods must be used (17).
[b] At 18°C, bleach for at least 10 min.

5.2 Fixing micro-injected *Drosophila* embryos

After imaging fluorescently tagged molecules in living embryos, it is also useful to fix the embryos and co-visualize the distribution of the labelled molecule with other cell components under improved imaging conditions. However, the standard methanol devitellinization procedure does not work with micro-injected embryos, requiring the removal of the vitelline membrane by hand. Below is my preferred method of removing the vitelline membrane, but several other variations have been described elsewhere (8).

Protocol 7. Fixation of micro-injected *Drosophila* embryos

Equipment and reagents

as in *Protocol 6*, plus the following:

- Glass Petri dish
- Glass Pasteur pipettes
- Glass scintillation vial
- Heptane

- PBS
- 3.7% formaldehyde in PBS, made from a 37% formaldehyde stock solution

Method

1. After micro-injecting and ageing the embryos appropriately (see *Protocol 6*), wash off the halocarbon oil by holding the coverslip with forceps at an angle above a glass Petri dish containing 10–20 ml of

Protocol 7. *Continued*

    heptane. Gently squirt heptane on to the embryos with a glass Pasteur
    pipette until they are dislodged from the glue and sink in the heptane.

2. Remove the embryos with a glass Pasteur pipette into a glass scintillation vial. Remove as much of the heptane (containing dissolved glue)
   as possible and replace with fresh heptane. Wash once more with
   heptane leaving 5 ml of heptane in the scintillation vial.

3. Add 5 ml of 3.7% formaldehyde in 1× PBS for 20 min if performing
   antibody staining. Alternatively, add 5 ml of 37% formaldehyde (the
   stock solution) for 2–5 min if performing mRNA *in situ* hybridizations
   or visualizing microtubules.[a]

4. Remove as much of the formaldehyde and heptane as possible. Wash
   twice in heptane. The following steps have to be performed rapidly to
   prevent the embryos drying out:

5. Pick all the embryos up in a small volume of heptane in a glass Pasteur
   pipette and squirt on to the back of a medium-sized sieve stuffed with
   a paper tissue to soak up the heptane.

6. Very rapidly (avoid embryos drying out completely) pick the embryos
   from the back of the sieve with a flap of double-sided sticky tape. Stick
   the tape on to a small plastic Petri dish (with the embryos facing
   upwards!) and press the tape down with forceps. Quickly (but gently)
   poor 1× PBS[b] on to the embryos. If necessary, the embryos can now
   be stored at 4°C for some time.

7. Remove the vitelline membrane by puncturing one side with a small
   needle and teasing the embryo out of the punctured side by pushing
   from the opposite side with a second needle. This can usually be
   achieved with minimal or no damage to the embryos. If the embryos
   burst or fall apart, then they have been insufficiently fixed.

[a] Fixation in 37% formaldehyde results in better morphological and microtubule preservation
but is incompatible with many antibody labelling procedures.
[b] It is important *not* to use detergents such as Tween-20 at this step.

5.3 Fixing GFP-expressing embryos

In non-embryonic tissues, visualizing GFP following fixation is fairly simple,
but over-fixation in formaldehyde abolishes the fluorescence. Strong fixation
with glutaraldehyde (for 20 min in 2.5% glutaraldehyde in 1× PBS) leads to
very strong fluorescence background, but light fixation with glutaraldehyde
(for 20 min in 0.05% glutaraldehyde in 1× PBS) is compatible with GFP
fluorescence. In contrast, GFP fluorescence is at least partly lost in embryos
upon fixation with formaldehyde, even when methanol is not used. It is unclear
what causes this problem, but paraformaldehyde and liquid formaldehyde

give similar results. GFP fluorescence is preserved in embryos if embryos are lightly fixed with glutaraldehyde, but hand-devitellinization is difficult with such embryos. Embryos fixed more heavily with glutaraldehyde (0.25% in $1\times$ PBS for 20 min) can be devitellinized by hand, but have a low level of background fluorescence which may interfere with imaging the GFP fluorescence. I have not yet found a satisfactory solution to these problems, but it is likely that better fixation conditions which preserve GFP fluorescence and allow devitellinization can be found.

5.4 Preparation, mounting and imaging living egg chambers

Egg chambers are particularly susceptible to morphological changes, developmental arrest and apoptosis induced by mistreating the females or incubating the egg chambers in aqueous media. The best method of preparing egg chambers for imaging is to dissect and mount them directly in halocarbon oil.

Protocol 8.   Mounting living *Drosophila* egg chambers

Equipment
- Etched tungsten micro-needles (500 μm) (Interfocus Ltd; 10130-20 and 26016-12)
- 24 mm × 50 mm coverslips (Laboratory Sales Ltd)
- PVC tape
- Series 95 halocarbon oil (Halocarbon Products Corporation)
- Breathable membrane (YSI Co. Inc., standard membrane kit, model 5793)

Method

1. Place newly eclosed females with males in well yeasted uncrowded bottles or vials for 2–3 days (at 25 °C). Treating the females well is very important, since oogenesis can become arrested in malnourished females, leading to abnormal ovaries.

2. Anaesthetize the females under CO_2 and transfer to a drop of halocarbon oil series 95.[a] Remove the ovaries directly into halocarbon oil.[a]

3. Place part of an ovary, in halocarbon oil, on to a coverslip with a strip of PVC tape with a 10 mm × 10 mm square cut out of its centre. Tease out individual ovarioles so that egg chambers are well separated. Note that fluid debris is inevitably created in the process of dissection, but should be kept to a minimum.

4. Place a 15 mm × 15 mm piece of breathable membrane on the oil to limit the drift of the egg chambers and extra fluid. This is particularly useful when attempting high-power imaging of living egg chambers (see *Figure 6*).

Protocol 8. *Continued*

5. Early staged egg chambers can only be imaged for approximately 2 h, after which they begin to look abnormal. Note that some imaging conditions can reduce the viability of the ovaries, for example, exposure to UV.

[a] Series 95 has the optimal viscosity for dissecting and mounting egg chambers. Lighter oil is easier for dissection but can leave the egg chambers exposed to drying as it spreads easily. In contrast, egg chambers are much more difficult to dissect in heavier halocarbon oils as they always clump together and are difficult to separate into individual ovarioles.
[b] Dissecting ovaries into 1× PBS, Ringer's or tissue culture medium all result in an instantaneous change in microtubule distribution in the oocyte.

5.5 Choice of expression vectors

A number of different methods of expressing GFP in embryos, egg chambers and imaginal discs have been used in *Drosophila*. In general, maternal GFP expression in the ovaries results in the earliest and easiest detection of GFP, overcoming the long delay in the appearance of fluorescence in the embryo (18). However, zygotic expression of GFP has also been possible, but there is a considerable delay between the appearance of the protein and the conversion of a sufficient proportion of the GFP to a fluorescent form. This conversion has been estimated to have a half-time of 4 h in the case of the wild-type protein in *Drosophila*. *Table 4* summarizes a number of constructs that have been used to express GFP fusion proteins in *Drosophila*.

5.6 Choice of GFP mutants

The different mutations vary in their excitation and emission wavelengths, speed of conversion to the fluorescent form, pH dependence and sensitivity to photobleaching. It is likely that double- or even triple-wavelength visual-

Table 4. Constructs used to express GFP in Drosophila

UAS-GAL4	UAS-GFPs65T	http://fly.ebi.ac.uk:7081/
	UAS-nlsGFP-S65T	(20)
	UAS-GFPnlslacZ	(21)
Heat-shock	hs-GFP-moesin	(22)
	hs-bcd-GFP	(23)
Other:	drosomycin-GFP	(24)
Maternal:	exuGFP	(25)
	PolyUb-nlsGFP and PUb-GFP	(18 and *Figure 6*)
	ncdGFP	(26)
	MaTub-stu-GFP and MaTub-mago	(27)
	bcd-GFP	(23)
Reviews:	(28–30)	(22)

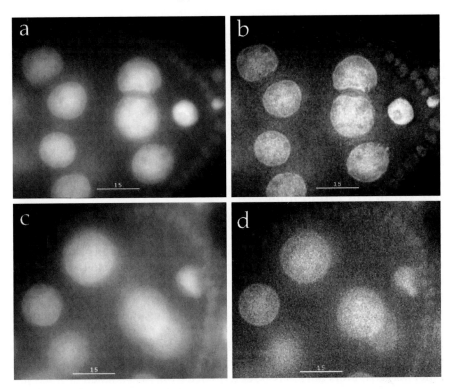

Figure 6. Examples of how the principles described in this chapter are used to improve imaging and deconvolution of *Drosophila* egg chambers. (a) Stage 7 egg chamber mounted in series 95 halocarbon oil with a membrane cover (see *Protocol 8*) and imaged with 1.534 immersion oil at a depth of approximately 20 μm from the coverslip. (b) The same image as in (a), after deconvolution. (c) Stage 7 egg chamber mounted in series 95 halocarbon oil without a membrane cover (see *Protocol 8*) and imaged with 1.512 immersion oil at a depth of approximately 20 μm from the coverslip. (d) The same image as in (c), after deconvolution. Note that the observed image has more out-of-focus light and that the deconvolution is much less effective, especially in the follicle cells which have a lower signal-to-noise ratio. The images were acquired using a Agard–Sedat deconvolution microscope (Applied Precision Inc.) with an Olympus IX70 inverted microscope, a 100x/NA 1.4 oil-immersion objective lens and a PXL CCD camera (Photometrics). The deconvolution was performed using DeltaVision software running on a Silicon Graphics computer, with 15 iterations. The images were contrasted equivalently so that a direct comparison can be made. The scale bars represent 15 μm.

ization of different GFP molecules will become routine in the near future. A detailed discussion of the properties of different GFP mutants is provided in Chapter 5. However, the utility of the different mutant forms of GFP depends critically on the organism. In bacteria and yeast cells in which the different mutants have been isolated and characterized, culturing conditions can be modified to change the fluorescence. For example, the emission wavelength of some GFP variants vary dramatically with pH (19). In *Drosophila* embryos or egg chambers, changing the culturing medium is unlikely to affect the conditions in the tissue itself. For this reason, the use of new GFP mutations in *Drosophila* embryos must be approached with caution. Wild-type GFP has been successfully used in *Drosophila*, but the S65T mutant form appears to offer an improvement in fluorescence intensity and reduced delay in the appearance of fluorescence. By the time this chapter is published, there will be many more GFP mutants and GFP transgenes in *Drosophila*. Currently, a side-by-side comparison of the properties of different GFP mutants and the accurate measurement of the differences in the delay in the appearance of fluorescence have not yet been performed in *Drosophila*.

5.7 Choice of filter cubes

There is a bewildering choice of different fluorescein isothiocyanate (FITC)- or GFP-specific filter cubes which can be used to visualize GFP fluorescence. In general, the choice of filter cube will depend on the organism and mutant GFP used, as well as the particular application. For example, different organisms have different degrees of autofluorescence; in some applications autofluorescence can be very useful, while in others it is detrimental. By the time this book is in print, there will be certainly more filter cubes and GFP mutants to choose from.

Most labs use either original Chroma filter cubes or Chroma filter cubes sold by the microscope manufacturer. The equivalent filter cubes made by other manufacturers can easily be determined from the parameters describing the excitation and emission filters as well as the dichroic mirror. Probably the best source of information on GFP filter cubes is Chroma itself, who publish booklets with reports by researchers who have tested various cubes. Their website (at http://www.chroma.com) contains considerable information on many of the filter cubes. I have found that FITC High Q filter no. 41001 is the best all-round filter cube for imaging GFP. This filter cube has an HQ480/40 excitation band-pass filter, a Q505LP dichroic mirror and an HQ535/50 band-pass emmision filter. In both yeast and *Drosophila* it gives the brightest signal and cuts out the most yellow background autofluorescence, resulting in the highest signal-to-noise ratio and overall brightness. Another very useful filter cube is a special manufacturing item from Olympus (U-MWIB/DIC-SP) which is able to simultaneously perform DIC imaging (in red light) with a polarizing dichroic mirror and GFP imaging. An analyser is not used with this filter cube so no GFP fluorescent light is lost.

6. Troubleshooting common imaging problems

There are many potential problems which can be encountered when imaging specimens especially at high power. *Table 5* summarizes many common imaging problems and their causes (also refer to the main text).

Table 5. Solving common imaging problems

Problem	Possible cause and solution
General	
Very poor z resolution	Check for spherial aberration, mount appropriately
Deconvolution artefacts	Check OTF is correct. Check spherical aberration and mounting
Asymmetric and pronounced airy rings	Correct spherical aberration with collar or immersion oil
Airy-rings have a segment taken out of them	Air bubble is present in the immersion oil
Different wavelengths are shifted	Chromatic aberration can be corrected by shifting the channels
Auto-fluorescence	Change antibody procedure or fixation to increase signal.
High Background	Change antibody procedure or fixation to reduce background.
Photo-bleaching	Use mountant with anti-fade reagent
Thermal drift of entire slide	Remove heat generating equipment and check air-conditioning.
Uneven fluorescence levels across the field	Check alignment of the microscope and light source
CCD—hot pixels or lines	Check grade of CCD and ensure CCD is being cooled correctly
CCD or confocal interference lines across image	Remove noisy power line or equipment
Living embroys	
Water droplets on surface	Desiccate the embryos for longer
Bright spots on surface	Wash embryos more extensively to remove yeast cells
Drift with respect to the coverslip or slide	Use stronger heptane-glue; stop embryo from detaching
Living egg chambers	
Drift with respect to the coverslip or slide	Use correct halocarbon oil and breathable membrane

Acknowledgements

I am grateful to Tony Shermoen and Carl Brown for discussions of various microscopy issues, and for allowing me to quote some of their measurements of the RIs of various media. I also thank Sabbi Lall, Nina Ehrenberg and Gavin Wilkie for their help in developing some of the embryo and egg chamber mounting techniques. All the investigators named above also made very valuable comments on this chapter.

References

1. White, N., Errington, R., Ericker, M. and Wood, J. (1996). *J. Microsc.*, **181**, 99.
2. Hiraoka, Y., Sedat, J. and Agard, D. (1990). *Biophys. J.*, **57**, 325.
3. Kiehart, D. P., Montague, R. A., Rickoll, W. L., Foard, D. and Thomas, G. H. (1994). In *Methods in Cell Biology*, Vol. 44: *Drosophila melanogaster: practical uses in cell and molecular biology* (ed. L. Goldstein and E. Fyrberg), p. 507. Academic Press, San Diego, CA.

4. White, R. (1998). In *Drosophila: A Practical Approach* (ed. D. Roberts), p. 215. Oxford University Press.
5. Wilkie, G. and Davis, I. (1998). *Elsevier Trends Journals Technical Tips Online* (an online reviewed journal, at http://www.biomednet.com/db/tto/), t01458.
6. Theurkauf, W. E. (1994). In *Methods in Cell Biology*, Vol. 44: *Drosophila melanogaster: practical uses in cell and molecular biology* (ed. L. Goldstein and E. Fryberg), p. 489. Academic Press, San Diego, C.A.
7. Patel, N. H. (1994). In *Methods in Cell Biology*, Vol. 44: *Drosophila melanogaster: practical uses in cell and molecular biology* (ed. L. Goldstein and E. Fryberg), p. 445. Academic Press, San Diego, C.A.
8. Wieschaus, E. and Nüsslein-Volhard, C. (1998). In *Drosophila: A Practical Approach* (ed. D. Roberts), p. 179. Oxford University Press.
9. Verheyen, E. and Cooley, L. (1994). In *Methods in Cell Biology*, Vol. 44: *Drosophila melanogaster: practical uses in cell and molecular biology* (ed. L. Goldstein and E. Fryberg), p. 545. Academic Press, San Diego, CA.
10. Pardue, M. L. (1994). In *Methods in Cell Biology*, Vol. 44: *Drosophila melanogaster: practical uses in cell and molecular biology* (ed. L. Goldstein and E. Fryberg), p. 333. Academic Press, San Diego, C.A.
11. Spector, D., Goldman, R. and Leinwand, L. (1998). *Cells: A Laboratory Handbook*. Cold Spring Harbor Laboratory Press, Cold Spring Harbor, NY.
12. Schrader, M., Hell, S. and van der Voort, H. (1996). *Appl. Phys. Lett.* **69**, **24**, 3644.
13. Kano, H., van der Voort, H., Schrader, M., van Kempen, G. and Hell, S. (1996). *BioImaging*, **4**, 187.
14. Jansson. (1970). *J. Opt. Soc. Am*, **50**, 596.
15. Agard, D. A. and Sedat, J. W. (1983). *Nature*, **302**, 676.
16. Roberts, D. and Standen, G. (1998). In *Drosophila: A Practical Approach* (ed. D. Roberts), p. 1. Oxford University Press.
17. Girdham, C. H. and O'Farrell, P. H. (1994). In *Methods in Cell Biology*, Vol. 44: *Drosophila melanogaster: practical uses in cell and molecular biology* (ed. L. Goldstein and E. Fryberg), p. 533. Academic Press, San Diego, C.A.
18. Davis, I., Girdham, C. H. and O'Farrell, P. H. (1995). *Dev. Biol.*, **170**, 726.
19. Miesenbock, G., De Angelis, D. and Rothman, J. (1998). *Nature*, **394**, 192.
20. Neufeld, T. P., DelaCruz, A. F. A., Johnston, L. A. and Edgar, B. A. (1998). *Cell*, **93**, 1183.
21. Shiga, Y., Tanakamatakatsu, M. and Hayashi, S. (1996). *Dev. Growth Differentiation*, **38**, 99.
22. Edwards, K. A., Demsky, M., Montague, R. A., Weymouth, N. and Kiehart, D. P. (1997). *Dev. Biol.*, **191**, 103.
23. Hazelrigg, T., Liu, N., Hong, Y. and Wang, S. (1998). *Dev. Biol.*, **199**, 245.
24. Ferrandon, D., Jung, A. C., Criqui, M. C., Lemaitre, B., Uttenweiler-Joseph, S., Michaut, L., Reichhart, J. M. and Hoffmann, J. A. (1998). *EMBO J.*, **17**, 1217.
25. Wang, S. X. and Hazelrigg, T. (1994). *Nature*, **369**, 400.
26. Endow, S. A. and Komma, D. J. (1996). *J. Cell Sci.*, **109**, 2429.
27. Micklem, D. R., Dasgupta, R., Elliott, H., Gergely, F., Davidson, C., Brand, A., Gonzalez-Reyes, A. and St Johnston, D. (1997). *Curr. Biol.*, **7**, 468.
28. Ludin, B. and Matus, A. (1998). *Trends Cell Biol.*, **8**, 72.
29. Brand, A. (1995). *Trends Genet.*, **11**, 324.
30. O'Kane, C. (1998). In *Drosophila: A Practical Approach* (ed. D. Roberts), p. 131. Oxford University Press.

7

Green fluorescent protein in plants

CHRIS HAWES, PETRA BOEVINK and IAN MOORE

1. Introduction

The introduction of the jellyfish green fluorescent protein (GFP) and its wavelength-shifted derivatives as reporter proteins for both gene expression and protein location has provided cell biologists with a powerful new tool with which to study cell structure and function *in vivo*. The protein has a number of intrinsic properties that make it attractive as a marker for cell biological studies. For instance, it is highly fluorescent and requires no co-factors or substrates other than oxygen to form the fluorochrome. It is relatively stable and pH insensitive, has a low level of cytotoxicity and in many instances does not interfere with the functioning of native proteins when expressed attached to them as chimeric constructs. Moreover, it has been genetically modified to alter its spectral and other properties such as folding rate, to make a family of tailor-made reporter proteins (1).

GFP chimeras were first used in plant cell biology to investigate the movement of viruses, by the construction of both fluorescent viruses and fluorescent viral proteins such as the movement proteins (2–4). Subsequently most major organelles have been successfully targeted including the nucleus (5), vacuole (6), mitochondria (7), the cell plate (8), Golgi apparatus and endoplasmic reticulum (9,10) and the cytoskeleton (11,12).

In this chapter we describe the various techniques available for the expression of GFP in plant cells. The reader should refer to Chapters 5 and 6 for detailed information about GFP and standard texts for information on the construction of suitable vectors, transformation of *Esherichia coli* and *Agrobacterium* and details of plant transformation protocols.

2. Choice of GFP

Since the first papers on the use of GFP as a reporter protein for gene expression in a variety of cell types (13), a number of reports have been published on the genetic modification of the protein. Most of these have been designed to make wild-type GFP a more efficient and useful tool for cell biologists, especially as a marker attached to heterologously expressed proteins. This re-engineering has been aimed at overcoming intrinsic problems with the native protein associated with low brightness, slow maturation of the chromophore, photo-isomerization and photobleaching of the protein (1). Therefore, it is advisable to make a careful choice of the GFP to be used before embarking on extensive transformation procedures.

Some of the first reports of the use of GFP in plants exploited viral expression systems such as potato virus X (2) and tobacco mosaic virus (14). This approach relies on the insertion of an extra sub-genomic promoter sequence that directs the synthesis of an inserted sub-genomic RNA, which acts as an mRNA for the introduced gene. Following this strategy, GFP can be expressed on its own, fused to a viral protein or as a chimera with another plant protein of interest. This strategy for GFP expression results in the production of high levels of fluorescent protein and has demonstrated the potential of GFP as a marker for viral infections and for various organelles (9,10). For this work it has proved possible to use unmodified wild-type *gfp* due to the high copy number and subsequent high levels of expression of the protein in infected cells.

Unfortunately, the first attempts to achieve stable expression of GFP in transgenic plants failed. Haseloff *et al.* (15) demonstrated that when wild-type GFP was expressed in *Arabidopsis* plants under the control of the cauliflower mosaic virus (CaMV) 35S promoter, fluorescence was not detectable. The problem was due to a deletion of 84 nucleotides in the GFP coding sequence that is recognized as a cryptic intron leading to aberrant mRNA splicing and production of a defective GFP. This problem was overcome by modification of the codon usage of the *gfp* gene to mutate the intron in order to permit correct expression of the gene to produce a wild-type protein product (15). As the cryptic intron contains sequences similar to those needed for the recognition of normal plant introns, it is likely that such modified GFPs will be routinely required for plant transformation work. A number of GFP variants have been produced with modified codon usage optimized for expression in mammalian cells (16) and are available from Clontech Labs (http://www.clontech.com). Co-incidentally, these modifications are predicted to abolish the mis-splicing and therefore should work in plant cells.

Although the above modifications provided high levels of GFP accumulation in transgenic plants, it transpired that such plants were frequently infertile (15). This problem can be overcome by targeting the protein into the endoplasmic reticulum (ER). This can be achieved by adding an appropriate N-terminal signal peptide to translocate the protein across the ER membrane and a C-terminal HDEL (His-Asp-Glu-Leu) sequence to maintain the protein in the ER lumen (mGFP4).

mGFP4 has been further modified by Val163→Ala and Ser175→Gly mutations, which prevent temperature-dependent mis-folding of the protein and thus give a more thermotolerant version. Also, and more important for plant work, an Ile167→Thr substitution has been used to enhance the amplitude of the 475 nm excitation peak of the GFP to a level equal to that of the 395 nm (UV) peak which predominates in the wild-type protein (17). This greatly increases the emission signal obtained from the protein when using argon ion laser lines and blue excitation for confocal microscopy. This new GFP variant, mGFP5er, therefore retains the advantages of the twin peak excitation of the protein, thus permitting the screening of material by hand-held long-wavelength UV lamps or with UV light in a fluorescence microscope. Such a modification has obvious advantages over many other optimized GFPs that are not detectable by UV light (18).

The other modification of GFP with great potential but yet to be fully exploited in plant biology, is engineering of the protein to change the excitation and emission wavelengths. Davis and Vierstra (19) have engineered a codon-modified GFP, smGFP, to be more soluble in the cytoplasm with excitation and emission spectra identical to those of the wild-type but brighter than the codon-modified form. This smGFP has been further modified to give a soluble blue-shift variant (emission shifted from 507 to 448 nm) and a soluble red-shift variant (excitation shifted from peaks at 397 and 480 nm to a single peak at 495 nm) producing blue fluorescent protein and red-shifted GFP. These proteins are available from the *Arabidopsis* Biological Resource Centre (http://aims.cps.msu.edu/aims). This strategy opens up the possibility of dual monitoring of gene expression and double transformations of plants targeting GFP with different spectral properties to different organelles giving a vital double-labelling system.

3. Expression systems

While many experiments may ultimately require the production of stable transformants, much useful information can be obtained through the use of transient expression systems. The production of transgenic plants expressing GFP can be achieved using standard transformation and gene expression technologies; these will not be discussed further in this chapter. Instead we describe a number of transient expression systems that have been successfully employed in the study of GFP targeting in plant cells.

3.1 Virus-mediated expression of GFP

Viruses replicate to high levels in plant cells and produce very large quantities of their structural proteins. They have, therefore, been exploited as expression vectors for plant systems, in the same way that baculoviruses have been used for protein expression in insect cells (20). Potato virus X (PVX) is one such plant virus, and has been used successfully to express the marker proteins GFP and GUS (2,21) and numerous other foreign proteins, often as fusions with the GFP (*Figure 1*) (9,10,22). The virus expression system has a

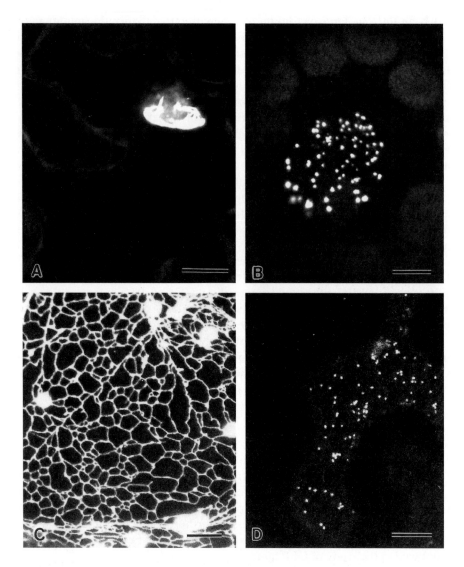

Figure 1. Confocal laser-scanning micrographs of virally expressed GFP in *Nicotiana* sp. leaf epidermal cells. (A) Aggregate of virus particles in a leaf cell. GFP was fused to the coat protein of potato virus X. Bar represents 20 μm. (B) GFP targeted to a field of plasmodesmata in an epidermal cell wall. GFP was fused to one of the virus triple-gene-block proteins thought to be involved in virus transport. Bar represents 5 μm. (C) Cortical endoplasmic reticulum (ER) in a leaf trichome. Free GFP was expressed with an N-terminal signal peptide and C-terminal KDEL sequence to promote ER retention. Bar represents 10 μm. (D) Golgi-targeted GFP. The signal anchor sequence (membrane-spanning domain plus cytoplasmic tail and a few lumenal amino acids) of a rat sialyl transferase was spliced to the N-terminus of GFP. The chimeric protein located to the individual Golgi stacks seen as bright dots in the micrograph. Bar represents 20 μm.

number of advantages over both the transient expression and the transgenic routes. Results are obtained very rapidly; 3–4 days after inoculating leaves of the host plant, with infectious transcripts of the viral vector constructs expressing GFP, lesions can be seen on the leaves under a long-wavelength (365 nm) UV lamp. The levels of protein expression are very high due to a combination of the strong viral promoter controlling the expression of the foreign protein, and the replication of the viral RNA providing more template for protein production. The viral infection is systemic so plants can be studied over extended periods of time. Another advantage is the avoidance of time-consuming transformation procedures. However, there are various limitations to the system that must be appreciated before undertaking an experimental programme. These include:

(a) Not all of the cells in the plant will be expressing the gene of interest as not all are infected with the virus.

(b) Some inserts may be unstable in the vector and the likelihood of instability increases with the size of the insert such that 3 kb may be close to the practical limit.

(c) Biochemical quantification of protein expression can be difficult because of differing levels of expression due to variations in levels of viral infection in the tissue.

(d) In targeting studies the frequently observed high levels of expression may promote mis-location of the fusion protein; however, this has not proved to be a problem in studies referenced here.

(e) As cells are infected with virus any pathogenic effects of the infection must be taken into account.

In the virus vector the coat protein sub-genomic promoter has been duplicated and a polycloning site created downstream for insertion of foreign genes. To create the vector pTXS.P3C2 from the full-length clone of PVX, the coat protein sub-genomic promoter was duplicated and a multiple cloning site was inserted between the duplicated sequences (*Figure 2*) (2). Foreign genes inserted into the multiple cloning site are therefore under the transcriptional control of the sub-genomic promoter immediately upstream of the site. A *Spe*I site at the 3{pri} end of the PVX sequence is used to linearize the plasmid before infectious run-off transcripts are made with T7 RNA polymerase; the T7 promoter is present immediately upstream of the PVX sequence.

Although PVX can only infect a limited number of species, various other plant viruses, such as tobacco mosaic virus (14), cowpea mosaic virus (23) and geminiviruses (24), have also been engineered to express GFP and other proteins. These various viruses differ in the levels of foreign gene expression, ease of manipulation of the viral genomes in cloning operations, and host specificities, but will offer the opportunity of expressing foreign proteins in a wide range of plants and in different tissues.

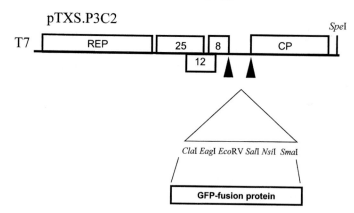

Figure 2. Insertion of GFP into the potato virus X vector pTXS.P3C2 to create pTXS-GFP. The viral cDNA is flanked by the T7 RNA promoter and *Spe*I recognition sequences. The open boxes represent coding sequences and lines represent untranslated sequence. The genes in the PVX vector are, from right to left: the replicase (REP), the triple gene block proteins with their approximate molecular weights in kDa and the coat protein (CP). The black triangles indicate a duplication of the pTXS.P3C2 coat protein sub-genomic promoter and a multiple cloning site. Modified GFP genes or chimeric GFP genes can be inserted in between these sites.

Protocol 1.   Preparation of viral transcripts

Reagents

- T7 RNA polymerase transcription kit (Ambion T7 'Message in Machine' kit)
- Standard molecular biology equipment

Method

1. Use standard cloning techniques (25) to clone the GFP of choice, GFP fusion or other protein of interest into the PVX vector. If appropriate restriction sites are not available in the DNA sequence to be inserted, PCR, with primers engineered to introduce suitable sites should be used. Ensure the gene is cloned in the correct orientation with respect to the promoter.

2. Prepare a high-quality, medium-scale, RNase-free preparation of the modified vector plasmid DNA.

3. To make infectious RNA, linearize the vector plasmid with *Spe*I or *Sph*I restriction enzymes. Use the linearized DNA as a template for the synthesis of capped transcripts, with a T7 RNA polymerase transcription kit following the manufacturer's instructions.

4. Lightly dust host plants, most commonly *Nicotiana benthamiana* and *N. clevelandii*[a] with aluminium oxide abrasive and inoculate with tran-

scripts (the transcripts are generally not cleaned or precipitated before inoculating). Transcript from approximately 0.2 μg of template DNA per leaf should give numerous (>10) lesions, depending on the construct. After inoculation, wash the abrasive off the leaves. Keep plants at approximately 23°C.

5. Three to four days after inoculation with a PVX-GFP virus, screen plants for small fluorescent lesions on inoculated leaves with a hand-held long-wavelength UV lamp.[b]

6. Leaf pieces can then be taken for observation of cells with fluorescence, but preferably, confocal microscopes.

[a] Note that if *N. tabacum* is used, viral infection is not systemic and observations are limited to infected leaves only.
[b] Some forms of GFP which have been optimized for imaging with blue excitation are not visible with hand-held UV lamps.

3.2 Transient expression

Transient expression systems are useful not only as relatively quick methods for checking the fidelity of plasmid construction but also for producing experimental material in their own right (see *Protocol 2*). For instance, transport of GFP to the non-acidic vacuolar compartment has been demonstrated in transiently expressing tobacco protoplasts (6). For transient expression GFP fusions should be cloned in a multi-copy plasmid under the control of a strong promoter such as CaMV 35S promoter. DNA must be of high quality and should be prepared either by caesium chloride gradient centrifugation or on an anion-exchange resin (for example Qiagen columns), followed by phenol/chloroform extraction and ethanol precipitation. Many different protocols exist for preparing protoplasts from a range of tissues. Here we give methods for the production of protoplasts from suspension cultures of *Arabidopsis thaliana* and introduction of plasmid DNA into protoplasts by high molecular weight polyethylene glycol (PEG) solutions (26) (see *Protocols 2* and *3*).

Protocol 2. Transient expression in *Arabidopsis*: preparation of protoplasts

Reagents

- Protoplasts are prepared from an *Arabidopsis* suspension culture (27) which should be sub-cultured weekly by diluting 1/10 in fresh medium, and should be used for protoplast preparation 5–6 days after sub-culture.
- W5/mannitol: W5 solution diluted 1/5 with 0.4 M mannitol
- Ma/Mg: 0.4 M mannitol, 15 mM MgCl$_2$, 1.5 mM Mes, pH 5.8

- Plasmolysis solution: 0.4 M mannitol, 3.0% sucrose, 8.0 mM CaCl$_2$, adjusted to pH 5.6 with KOH
- Enzyme solution: 1.00% cellulase (Onozuka R10) 0.25% Macerozyme (Onozuka R10) in plasmolysis solution
- W5 solution: 5.0 mM glucose, 154 mM NaCl, 125 mM CaCl$_2$, 5.0 mM KCl, 1.5 mM Mes, adjusted to pH 5.6 with KOH

Protocol 2. *Continued*

Method

1. Collect 20 ml of 6 day old *Arabidopsis thaliana* suspension culture cells in Falcon tubes or similar.

2. Lightly pellet (100*g*, 10 min) and re-suspend the cells in 40 ml of plasmolysis solution and incubate at room temperature for 30 min.

3. Re-pellet cells as described above and incubate in 40 ml of wall-digesting enzymes in 50 ml Falcon tubes, at room temperature in the dark for 2 h, then for 30 min with light shaking on a rocking table.

4. Pellet protoplasts by centrifugation for 5 min at 100*g* and wash by resuspension in 30 ml W5/mannitol followed by re-centrifugation.

5. Wash twice with 10 ml Ma/Mg and resuspend in 6 ml Ma/Mg.

Protocol 3. Transient expression in *Arabidopsis*: DNA preparation and PEG-mediated transformation

Reagents

- PEG solution: 0.4 M mannitol, 0.1 M Ca(NO$_3$)$_2$, 40% PEG 4000 (BDH Analar)
- Protoplast culture medium: Murashige and Skoog medium (Sigma) supplemented with 0.4 M sucrose, 250 mg/L xylose, adjusted to pH 5.8 with 0.1 M KOH

- Herring-sperm DNA (Sigma), 10–20 µg DNA/µl in water. This must be extensively purified by repeated phenol extraction and precipitation prior to use in transformation.
- W5/mannitol (see *Protocol 2*)

Method

1. Mix 250 µg herring-sperm carrier DNA with 10–50 µg super-coiled recombinant DNA.

2. Sterilize with the addition of 25 µL of chloroform.

3. While the protoplasts are being washed, vortex the DNA/chloroform mixture and centrifuge (1400*g*, 5 min) to separate the DNA from the chloroform.

4. Pipette each DNA sample into the periphery of separate sterile Petri dishes and replace the lids, taking care not to transfer the lower chloroform phase.

5. When the protoplasts are ready, pipette 350 µl of PEG solution into the periphery of each Petri dish and a 300 µl drop of protoplast suspension into the centre of each dish.

6. Take up the DNA from the periphery of each dish and gently but thoroughly mix it into the protoplast droplet by smooth movement of the pipette tip as the DNA is expelled.

7. Immediately after adding the DNA, add the PEG and mix as above. Mixing of DNA and PEG with protoplasts should take between 30 and 60 sec.

8. Slowly dilute the incubation mixture with W5/mannitol (see *Protocol 2*) by adding 500 μl, 1 ml, 2 ml and 4 ml at 15 min intervals.

9. Collect protoplasts into a 15 ml sterile Falcon tube and allow to settle (10–15 min).

10. Add 2 ml of protoplast culture medium but do not attempt to re-suspend the pellet.

11. Incubate protoplasts for 24–36 h in the dark at 20°C.

12. Screen for GFP-expressing protoplasts with a fluorescence microscope.[a]

[a] If a GFP with a UV excitation peak is used, screening with a UV filter set is often preferable to screening with a blue excitation (or FITC) filter set.

Although transient expression in protoplasts is a useful technique for validating the fidelity of GFP chimeras the subsequent analysis of GFP location can present various problems. For instance, the number of protoplasts expressing detectable levels of the protein may be limited to a few percent; not all cells receive equal quantities of DNA so expression levels are difficult to control and spherical cells such as protoplasts are not ideal specimens for microscopical analysis. This almost certainly requires confocal microscopy for observing GFP at the organellar level.

Another efficient method for transiently expressing GFP constructs in plant tissues is 'biolistic' gene bombardment (4,11,12), leaves and suspension cultures being particularly suitable material for this technique. With this technique plant material is bombarded with small (1 μm) gold particles coated with DNA. The advantage of this method over protoplast expression is that expression in walled cells can be studied, and data can be collected both from the epidermis and underlying mesophyll cells.

3.3 *Agrobacterium*-mediated transient expression in *Nicotiana* sp.

An extremely quick and efficient method for expressing GFP constructs in plants is to infiltrate intracellular spaces of leaves with a suspension of *Agrobacterium tumefaciens* that carries the GFP construct in a suitable binary vector (*Figure 3*). This procedure evolved from the demonstration of *A. tumefaciens*-mediated transient expression in plant cells (28). A culture of *Agrobacterium* is resuspended in an infiltration medium and simply pressure-injected though the stomata of the abaxial leaf epidermis. After about 2 days, GFP expression can be monitored by a hand-held UV lamp (if the GFP used

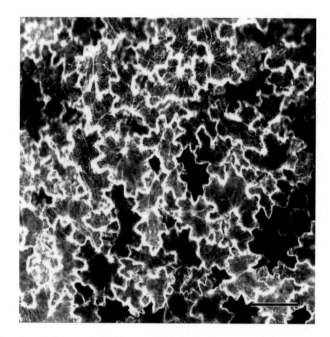

Figure 3. *Agrobacterium tumefaciens*-mediated transient expression of GFP in *Nicotiana clevelandii* leaf epidermal cells. This confocal micrograph shows the high level of expression, over large areas of leaf, that can be obtained in the endoplasmic reticulum with a signal peptide–GFP–HDEL construct. Bar represents 100 μm.

has a UV excitation peak) and areas expressing the protein can be excised for microscopy. This technique gives high expression levels of GFP constructs, which may fade about 4–5 days after inoculation of the leaves.

A number of considerations have to be made in the design of such an experiment:

(a) In our experiments we use binary vectors designed to maximize transcription and translation of GFP in plants (*Figure 4*). However, the high expression levels achieved suggest that standard CaMV 35S-based expression vectors may well be adequate.

(b) It is important to ensure that the GFP is not detectably expressed in the bacteria. With other reporter systems this problem has been overcome by insertion of a plant intron into the reporter gene. Unfortunately such intron-containing GFPs have yet to be constructed. However, we have found that a variety of GFP fusions expressed from vector pVKH18En6-mGFP5ER (*Figure 4*) show no detectable bacterial fluorescence. This may result from the strong *lac* promoter, downstream of the expression cassette, which may produce an anti-sense transcript.

pVKH18En6-mGFP5ER

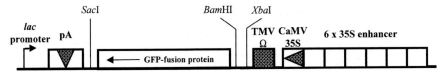

Figure 4. Shown schematically is a portion of the T-DNA of pVKH18En6-mGFP5ER. This carries the mGFP5 protein, which is targeted to and accumulates within the ER of plant cells. It is expressed from a CaMV promoter containing six copies of the enhancer in tandem array (30). The 5{pri} untranslated leader sequence includes the Ω translational enhancer from tobacco mosaic virus (TMV). pA indicates the polyadenylation signal of the *A. tumefaciens* nopaline synthase gene. A *lac* promoter located downstream of the polyadenylation signal may generate an anti-sense transcript in bacteria, helping to suppress GFP expression in *A. tumefaciens*. The expression cassette shown here is derived from pE6113GUS (30) and is located near the T-DNA right border in pVKH18En6-mGFP5ER. This plasmid also carries a hygromycin resistance marker near the T-DNA left border, and this can be used to select transgenic plants.

(c) The protocol given below utilizes a bacterial suspension that is likely to provide high levels of expression. As high expression levels may result in mis-targeting, it is advised that serial dilutions of bacteria are used to optimize expression levels.

Protocol 4. *Agrobacterium*-mediated expression of GFP

Reagents
- Healthy plants of *Nicotiana tabacum, N. clevelandii* or *N. benthamiana*
- *Agrobacterium* YEB medium: 0.5% (w/v) beef extract, 0.1% (w/v) yeast extract, 0.5% (w/v) peptone, 0.5% (w/v) sucrose, 2 mM MgSO$_4$.7H$_2$O
- Infiltration medium: 50 mM Mes pH 5.6, 0.5% (w/v) glucose, 2 mM Na$_3$PO$_4$, 100 μM acetosyringone (Aldrich) from 200 mM stock in DMSO

Method

1. Use standard cloning techniques to clone the plant-modified GFP construct of choice into *Agrobacterium tumefaciens*.

2. Culture the *Agrobacterium* in YEB medium at 28°C to stationary phase (24–48 h).

3. Take 1.0 mL of the culture, pellet, wash once in infiltration medium and resuspend to an OD$_{600}$ of 0.5–0.6.

4. Inject bacterial suspension into the abaxial epidermis of plant leaves from a 1 ml plastic syringe by pressing the nozzle against the leaf surface (do not use a needle). Spread of liquid entering the leaf via stomata can be visualized by a darkening of the leaf. The boundaries of the infiltrated area may be marked with an indelible pen.

Protocol 4. *Continued*

5. Incubate plants for 2 days at 20–25 °C.

6. Check for expression in leaves with a hand-held long-wavelength UV lamp after 2 days if a GFP with a UV excitation peak has been used.

7. If fluorescent patches can be observed, excise portions of the leaf for microscopical analysis, otherwise excise a piece of leaf from within the marked area.

This method offers a number of advantages over protoplast-based protocols:

(a) Within the infected area most if not all cells appear to express the construct with relatively little variability between cells. This includes both the abaxial and adaxial epidermis, guard cells, palisade and spongy mesophylls, but rarely trichomes and vascular tissue.

(b) Rossi *et al.* (28) have shown that expression from the T-DNA varies linearly with bacterial concentration over three orders of magnitude, offering the opportunity to control GFP expression levels.

(c) Intact cells and tissues can be studied with minimal disruption, although it is not clear to what extent the bacterial infection alters cell physiology.

(d) The geometry of epidermal cells, which have large vacuoles and a thin layer of cytoplasm immediately below the cuticular wall, greatly facilitates observation by both conventional epifluorescence microscopy and confocal imaging. For example, we have been able to distinguish the distribution of ER- and Golgi-targeted GFP using a conventional epifluorescence microscope.

4. Double labelling

Although the obvious advantage of GFP is that it can be imaged in living cells, occasions arise where it is desirable to fluorescently label other structures within GFP-expressing cells. Fortunately, GFP will remain fluorescent after both paraformaldehyde fixation and light glutaraldehyde fixation, thus permitting the use of probes such as rhodamine-conjugated phalloidin (10) or standard immunofluorescence protocols with red or far-red emitting fluorochromes. GFP will also retain antigenicity after fixation and preparation for immunogold labelling using the progressive lowering of temperature technique and embedding in acrylic resins (10,29). Thus, it is possible to confirm the location of GFP fusion proteins at the ultrastructural level, which can be important when proteins are targeted to some of the small cellular organelles such as mitochondria, plastids and Golgi. We have encountered difficulties in obtaining GFP antibodies that are suitable for immunogold labelling of plant tissue. Consequently, it may be advisable to incorporate one of the widely used epitope tags into the GFP fusion if at some stage in the future immunogold labelling is envisaged.

5. Microscopy

One of the major advantages of GFP when used to tag proteins of interest is that cellular events can be studied *in vivo in planta*. This requires the mounting of either whole plantlets, as in the case of *Arabidopsis* seedlings, or relatively large pieces of tissue, such as leaf segments, roots tips, etc. It is important to remember that GFP fluorescence will not develop in an anoxic environment so that microscope preparations must not be sealed with an impermeable medium such as nail varnish. In most instances it is sufficient to mount the material in a drop of water and cover with a long coverslip which can be held in position with strips of adhesive tape at either end.

Single cells, such as transformed suspension cultures or protoplasts used for transient expression, can be monitored by conventional epifluorescence microscopy. However, with tissue samples it may be necessary to use either confocal microscopy or conventional fluorescence combined with low-light-level cameras and deconvolution software (Chapters 4–6). However, with the observation of living plant material specific problems such as autofluorescence of chlorophyll and rapid cytoplasmic streaming or organelle movements are often encountered which can be overcome by selective use of filter sets and optimizing microscope settings (*Table 1*). It is worth remembering that GFP is relatively resistant to photobleaching. Therefore, for image capture with confocal microscopes high laser power can be used if necessary. When

---

Table 1. Optimizing the final image in GFP-expressing plant material by confocal microscopy

Problem	Action	Result
Autofluorescence of chlorophyll	Use a narrow band emission filter set, e.g. 510–525 nm.	Reduction in autofluorescence from chloroplasts, but with some overall reduction in signal which may have to be compensated for;
Low signal from GFP that cannot be compensated for by change in image brightness contrast settings	Increase laser intensity and/or increase pinhole aperture.	increase in signal from GFP tagged proteins or organelles, reduction in and confocality and resolution
Blurred image due to cytoplasmic movements	Increase laser intensity and/or increase pinhole aperture. Reduce image averaging and time of single line scan. If necessary, collect data from a defined region of interest.	Reduction in resolution and confocality; quick collection of time-resolved data sets for presentation as movies

---

combined with a wide pinhole aperture good quality images can be obtained from single line scans without having to resort to image averaging techniques. Although this strategy reduces confocality and the ultimate resolution of the image, it can help overcome problems of imaging blurring owing to organelle movement during the image capture process.

References

1. Cubitt, A.B., Heim, R., Adams, S.R., Boyd, A.E., Gross, L.A. and Tsien, R.Y. (1995). *Trends in Biochemical Sciences* **20**, 448.
2. Baulcombe, D.C., Chapman, S.N. and Santa Cruz, S. (1995). *Plant Journal* **7**, 1045.
3. Oparka, K.J., Roberts, A.G., Prior, D.A.M., Baulcombe, D. and Santa Cruz, S. (1995). *Protoplasma* **189**, 133.
4. Itaya, A., Hickman, H., Bao, Y.M., Nelson, R., and Ding, B. (1997). *Plant Journal* **12**, 1223.
5. Grebenok, R.J., Pierson, E., Lambert, G.M., Gong, F.-C., Afonso, C.L., Haldeman-Cahill, R., Carrington, J.C. and Galbraith, D.W. (1997). *Plant Journal* **11**, 573.
6. Di Sansebastiano, G.-P., Paris, N., Marc-Martin, S. and Neuhaus, J.-M. (1998). *Plant Journal* **15**, 449.
7. Köhler, R.H., Zipfel, W.R., Webb, W.W. and Hanson, M.R. (1997). *Plant Journal* **11**, 613.
8. Gu, X. and Verma, D.P.S. (1997). *Plant Cell* **9**, 157.
9. Boevink, P., Santa Cruz, S., Hawes, C., Harris, N. and Oparka, K.J. (1996). *Plant Journal* **10**, 935.
10. Boevink, P., Oparka, K., Santa Cruz, S., Martin, B., Betteridge, A. and Hawes, C. (1998). *Plant Journal* **15**, 441.
11. Kost, B., Spielhofer, P. and Chua, N.-H. (1998). *Plant Journal* **16**, 393.
12. Marc, J., Granger, C.L., Brincat, J., Fisher, D.D., Kao, T.-H., McCubbin, A.G. and Cyr, R.J. (1998). *Plant Cell* **10**, 1927.
13. Chalfie, M., Tu, Y., Euskirchen, G., Ward, W.W. and Prasher, D.C. (1994). *Science* **236**, 802.
14. Heinlein, M., Epel, B.L., Padgett, H.S. and Beachy, R.N. (1995). *Science* **270**, 1983.
15. Haseloff, J., Siemering, K.R., Prasher, D.C. and Hodge, S. (1997). *Proceedings of the National Academy of Sciences of the USA* **94**, 2122.
16. Haas, J. (1996). *Current Biology* **6**, 315.
17. Siemering, K.R., Golbik, R., Sever, R. and Haseloff, J. (1996). *Current Biology* **6**, 1653.
18. Heim, R. and Tsien, R.Y. (1996). *Current Biology* **6**, 178.
19. Davis, S.J. and Vierstra, R.D. (1998). *Plant Molecular Biology,* **36**, 521.
20. King, L.A. and Posse, R.D. (1992). *The Baculovirus Expression System. A Laboratory Guide.* Chapman Hall, London.
21. Chapman, S.N., Kavanagh, T. and Baulcombe, D.C. (1992). *Plant Journal* **7**, 1045.
22. Blackman, L.M., Boevink, P., Santa Cruz, S., Palukaitis, P. and Oparka, K.J. (1998). *Plant Cell* **10**, 525.
23. Verver, J., Wellink, J., VanLent, J., Gopinath, K. and Vankammen, A. (1998). *Virology* **242**, 22.

24. Sudarshana, M.R., Wang, H.L., Lucas, W.J. and Gilbertson, R.L. (1998). *Molecular Plant–Microbe Interactions* **11**, 277.
25. Sambrook, J., Fritsch, E.F. and Maniatis, T. (1989). *Molecular Cloning: A Laboratory Manual*, 2nd edn. Cold Spring Harbor Laboratory Press, Cold Spring Harbor, NY.
26. Abel, S. and Theologis, A. (1994). *Plant Journal* **5**, 421.
27. May, M. and Leaver, C.J. (1993). *Plant Physiology* **103**, 621.
28. Rossi, L., Escudero, J., Hohn, B and Tinland, B (1993). *Plant Molecular Biology Reporter* **11**, 220.
29. VandenBosch, K.A. (1991). In *Electron Microscopy of Plant Cells* (ed. J.L. Hall and C. Hawes), p. 181. Academic Press, London.
30. Mitsuhara, I. *et al.* (1996). *Plant Cell Physiology* **37**, 49.

Recently fluorescence resonance energy transfer (FRET) microscopy using fluorescent proteins has been demonstrated in root hairs using a calcium sensitive yellow-cameleon-2 construct. This consists of cyan (CFP) and yellow (YFP) derivatives of GFP linked by calmodulin and the M13 calmodulin-binding protein.

Gadella, T.W.J., van der Krogt, G.N.M. and Bisseling, T. (1999). *Trends in Plant Science*, **4**, 287–291.

Fluorescence microscopy in yeast

IAIN M. HAGAN and KATHRYN R. AYSCOUGH

1. Introduction

The reductionist desire of the modern biologist to analyse complex problems in the genetically amenable systems which offer the fewest variables has seen yeasts come to the fore in modern cell biology as the leading model eukaryotes. Both the budding yeast *Saccharomyces cerevisiae* and the fission yeast *Schizosaccharomyces pombe* offer many advantages for the analysis of diverse aspects of cell biology. They are unicellular but also have limited options of differentiation in sexual reproduction and pseudohyphal growth. Their usual mode of growth and the tight correlation between growth and cell division have made them excellent systems for the analysis of the control of cell cycle progression. Moreover the strong classical and molecular genetics have meant that yeasts are not only a favourite of the research scientist but also appeal to the biotechnologist in their desire to produce large quantities of cloned proteins. These factors, combined with its small genome size, made budding yeast the obvious choice for a concerted effort to establish the first complete DNA sequence of a eukaryote (see *Table 1* for genome website). This information in turn has heightened interest in yeast as a model system as many genes can be identified *in silico* and the boundaries of potential complexity of different pathways have been set. It is now known exactly how many members of a particular protein family exist in *S. cerevisiae*. This defines limits for modelling a particular process as the possibility that perhaps another isoform of a particular protein exists is no longer an acceptable escape clause to consolidate a flagging model. The combination of the information provided by the genome and the technological advances in mass spectroscopy mean that it is now possible to identify a protein on the basis of a fraction of the information that is required in virtually every other system. The exact mass of a peptide in a mass spectroscope can be compared against all possible combinations in the genome sequence. Information from four distinct peptides from one molecule is sufficient to categorically identify the protein from which the peptides were derived. Nowhere has the potential of these combined technological advances been more appropriately demonstrated than with the

Table 1. Useful websites for information on yeast and fluorescence techniques

Website address	Notes
http://genome-www.stanford.edu/Saccharomyces/ yeastlabs.html	links to many *S. cerevisiae* labs
http://www.sacs.ucsf.edu/home/HerskowitzLab/	well maintained, useful *S. cerevisiae* protocols
http://genome-www.stanford.edu/group/botlab	well maintained, useful *S. cerevisiae* protocols
http://flosun.salk.edu/~forsburg	well maintained useful *S. pombe* reference site
http://www2.bio.uva.nl/pombe/	well maintained useful *S. pombe* reference site
http://www.bio.uva.nl/pombe/handbook/	many *S. pombe* protocols
http://www.clontech.com/clontech/Manuals/GFP/TOC. html	useful site for GFP information
http://www.probes.com	Molecular Probes site: useful information on fluorophores, secondary antibodies and vital stains
http://www.invitrogen.com/genestorm/	InVitrogen site for His-tagged *S. cerevisiae* genes
http://genome-www.stanford.edu/Saccharomyces/	*S. cerevisiae* genome site
http://www.sanger.ac.uk/Projects/S_pombe/	*S. pombe* genome site

To subscribe to the yeast newsgroup, send the message 'SUBSCRIBE bionet-news.bionet.molbio. yeast' to the address MXT@dl.ac.uk. To unsubscribe, send the message 'UNSUB bionet-news.bionet. molbio.yeast' to the same address. Subscribers receive messages/enquiries daily about yeast-related topics. A message can be sent to all other subscribers using the address yeast@dl.ac.uk.

recent identification of 11 new genes as being those that encode components of the budding yeast spindle pole body (SPB) (1).

The intense interest in *S. cerevisiae* is heightening interest in *S. pombe* for much the same reasons and it is becoming apparent, from the limited sequence analysis available at the time of press (around 70% complete), that at least 20% of the *S. pombe* sequences identified in the genome sequencing project to date have no homologue in *S. cerevisiae* (see *Table 1* for website). This in turn is increasing the interest of a diverse community in the complexities of the cell biology of both yeasts.

We have therefore attempted to cover a number of protocols that are in common use for the staining of both budding and fission yeast for light microscopy. The vast majority of this chapter is devoted to the complexities of immunofluorescence protein localization. We will not cover the equally complex technology for DNA localization through fluorescence *in situ* hybridisation (FISH) as this has been recently reviewed in a similar volume (2) and in chapter 3. After an analysis of each stage of the procedure we present a number of standard protocols for immunofluorescence which can act as starting points for the localization of a protein of interest. The protocols, references and website links provided here should provide a framework around which to develop strategies to localize a molecule of choice.

2. Practical considerations when working with yeast

2.1 Difficulties encountered when working with yeast

There are a number of problems associated with protein localization in fungal systems such as the yeasts which are rarely encountered in the more commonly studied higher organisms. One of the most obvious differences is the presence of a cell wall. This generates a barrier to fixation and perturbs some of the standard manipulations that are routinely used with many other cell types. The second feature of fungal systems that affects immunolocalization technology is their very dense cytoplasm. This presents two problems: firstly, the fixative must penetrate this dense matrix and, secondly, there must be enough room for the antibodies to diffuse for specific labelling.

While it is possible to talk in such general terms of the cytology of fungi, the processing of different fungi for immunofluorescence microscopy varies almost as much within the field as it does in comparison with higher systems. Thus, techniques that preserve microtubules perfectly in *S. cerevisiae* (3) do not preserve the full array of *S. pombe* microutubules (4).

It should be remembered that far more time and energy have been investigated in protein localization in *S. cerevisiae* than *S. pombe*. Thus, many of the techniques and tricks described for *S. cerevisiae* have simply never been tried on fission yeast and so researchers in the fission yeast community would do well to heed the investment and experience of the *S. cerevisiae* community in adapting and developing new technologies.

2.2 Growing yeast

The growth state of the cells and the type of medium in which they are cultured can both have considerable bearing upon the quality of fixation for immunofluorescence. Although this effect is often negligible, rich media generally seem to give more reliable and better fixation than minimal media. If minimal medium must be used because selection is required to maintain a plasmid, cells can be pelleted from the selective medium and the cells resuspended in rich medium prior to fixation (5).

Growth in low-glucose minimal media, which reduces the carbohydrate available for cell wall synthesis, thus reducing the barrier to aldehyde penetration, also improves fixation, but can affect phenotypic expression (6). Osmotic buffering of a medium through the addition of sorbitol can have dramatic effects on the quality of fixation. For example, it is not normally possible to preserve cytoplasmic microtubules in fission yeast by 3% formaldehyde, but these microtubules are well preserved if an equal volume of pre-warmed medium which is 2.4 M with respect to sorbitol is added 5 min before fixation (7–9). It should also be noted that the particular stage of the life cycle may also influence staining. It is generally more difficult to stain fission yeast cells during sexual differentiation than the same strain with the same antibody during vegetative growth. It is not clear whether this effect is due to the medium, the considerable increase in non-specific proteolysis that accompanies sexual differentiation or the altered cell wall structure of differentiating cells.

2.3 Harvesting and fixing cells

2.3.1 Considerations when fixing yeast cells

The aim of fixation is to permanently 'freeze' cell structure in the *in vivo* state. Generally this is achieved either by rapid dehydration, to induce precipitation of proteins, or by the use of a molecular 'glue' to cross-link the cells components. If immunofluorescence is to be carried out after fixation it is important that the specific antigen is preserved in its native state in order for an antibody to recognize it. The chemistry of fixation by cross-linking is complex but essentially it involves chemical reactions of the fixative with specific amino acids on different proteins. If this amino acid forms part of the epitope one wishes to see, then immunofluorescence will be difficult. Another problem arises if the antigen is obscured by being buried deep within a complex. In this context, the strongest fixation is sometimes not the best way to see the protein of choice, as epitopes may be more exposed on a weakly fixed sample that is loosely held in place than on well fixed samples in which structural integrity is totally preserved. There are inherent advantages and disadvantages with the use of cross-linking or precipitation in yeast immunofluorescence, so each approach will be addressed in turn.

2.3.2 Chemical fixation to cross-link proteins

This is the most popular method for fixation in use in yeast cytology at present. Formaldehyde is the reagent of choice, either with or without supplementary glutaraldehyde. Generally most budding yeast researchers buy ready-made formaldehyde solutions whereas for fission yeast it appears to be critical for most fixations to make up fresh aldehyde on the day of use.

Because the polymer content of aldehyde solutions increases with time, ultra-pure fresh glutaraldehyde is generally used for electron microscopy to avoid artefacts during fixation. Experience has shown, however, that some of the best immunofluorescence results often come with less pure solutions which have been stored at 4 °C for up to 3 years. This may reflect the greater need for longer-range but weaker fixation for fluorescence microscopy rather than the preservation of fine detail which is critical for the types of electron microscopy in which glutaraldehyde is used. It should, however, be noted that glutaraldehyde solutions do eventually go off and the solution should be replaced when there is a consistent drop in the quality of fixation.

One of the problems experienced with aldehyde fixation is that of penetration of the fixative into the sample; the fixative needs to cross-link efficiently but not so efficiently that it impedes the penetration of further aldehyde. In situations where strong fixation is of paramount importance, such as in the visualization of microtubules in fission yeast, the problem is overcome through the use of both glutaraldehyde and formaldehyde fixatives. The theory is that while formaldehyde penetrates rapidly, it is a weaker fixative than the more slowly penetrating glutaraldehyde. Thus, the formaldehyde maintains cell structure until the glutaraldehyde completes the fixation to the desired level. If glutaraldehyde alone were to be used, the outside of the cell would be beautifully fixed while the nucleus would start to lose structures. Thus, optimum results are often obtained if the glutaraldehyde is generally added 30 sec or so after the formaldehyde. One problem encountered with the use of glutaraldehyde is that high concentrations result in non-specific, background fluorescence. Therefore the lowest concentration that still preserves the specific structure should be used (e.g. to see microtubules in fission yeast 0.2% glutaraldehyde is added).

It should be stressed that glutaraldehyde does destroy the antigenicity of many antigens. Fission yeast cells stain well with the monoclonal antibody N350 anti-actin (Amersham) if they are fixed with 4% formaldehyde; however, if 0.2% glutaraldehyde is used in addition to the 4% formaldehyde, staining is abolished.

2.3.3 Quenching
Because glutaraldehyde is a di-aldehyde, there are many free aldehyde groups available to react with the incoming antibodies and so increase background fluorescence. These can be reduced by a 5 min wash with 1 mg/ml sodium borohydride in solution. Such solutions should be made up fresh for each wash just before use.

2.3.4 The duration of chemical fixation is a key consideration
The time that a cell spends in fixative is also a crucial factor in protein localization. For many fusions with the haemagglutinin (HA) epitope, staining is completely abolished by incubations in fixative for more than 10 min. Similarly the staining efficiency and distribution of the SPB antigen Spc110p diminishes with the duration of fixation (10). A related and real concern of fixation-dependent distribution effects have been examined extensively and systematically by Melan and Sluder (11). In their systematic study they have highlighted the real and often overlooked dangers of fixation artefacts during sample preparation. As noted above, some of these artefacts can be overcome by the inclusion of an osmotic stabilizer in the buffer for fixation.

2.3.5 Fixing by protein precipitation: solvent fixation
Cold solvents rapidly dehydrate samples, resulting in the precipitation of the proteins *in situ*. Solvent fixation has the disadvantage that cells shrink as they

are dehydrated but this has been put to good use in several protocols as it can help expose epitopes that would be hidden by chemical fixation (12). Furthermore, because the epitope is not modified by reaction with the fixative, antigen preservation is generally much better during solvent rather than aldehyde fixation. The most commonly used budding yeast protocols use a solvent fixation after an initial chemical fixation with formaldehyde (see *Protocol 2*). This second step probably helps to open up the fixed cell structure to the antibody.

One problem that is associated with simple solvent fixation is that the cytoplasm is generally excellently preserved, while nuclear preservation is highly variable. This may be explained by the fixation being insufficiently rapid to preserve the central structures before their order starts to break down because they are the last regions to feel the effect of the solvent. The more central a structure, the more chance it has of being destroyed by the other changes in the cell before the protein in this region is precipitated. Thus, more cytoplasmic microtubules are preserved by solvent fixation in fission yeast than with combined aldehyde fixation, but the latter routinely gives better spindle structure (4). One way around the problem of loss of structure in the middle of the cell is to put make fixation virtually instantaneous by putting very small samples into extremely cold liquids such as liquid helium or liquid propane and then dehydrating the frozen samples with solvent. This technique is so specialized, however, that it is not recommended for routine fluorescence microscopy, rather for electron microscopy where precise preservation is a necessity (13). The temperature of the solvent for fixation for immunofluorescence is important as solvents at $-80\,^{\circ}C$ give much better preservation of the mitotic spindle than the same solvent at $-20\,^{\circ}C$.

Having raised these caveats, it must be added that perfect preservation is not always necessary for a particular experiment: sometimes simple data, such as determining whether a protein is cytoplasmic or nuclear, can be informative.

2.3.6 Considerations when harvesting cells

It has been noted in *S. cerevisiae* that harvesting cells by centrifugation can disrupt the actin cytoskeleton and organelle morphology (14,15). If chemical fixation is used this is not generally a problem, as cells are directly fixed in culture by the rapid addition of 0.1 vol. of stock solution. For solvent fixation, however, fixing in culture is not practical and cells are harvested by filtration on to glass-fibre filter pads. With filtration it is important not to harvest too many cells at once, as the filter quickly gets clogged and the pressure on the cells escalates, raising concerns about the maintenance of their structural integrity and their physiology. Thirty millilitres of a log-phase culture is around the maximum that should be filtered before harvesting is slowed down on 24 mm filter pads. Fixation is achieved by simple plunging of the filter into cold solvent in a tube. Because the filter warms up the solvent, and because temperature affects the fidelity of solvent fixation, the quality of fixation may be reduced if more than one filter is put into a tube. It is therefore advisable to use a single tube per sample, use a large volume of solvent, and use the solvent at as low a temperature as possible, i.e. at –80°C rather than –20°C. A final form of harvesting is to scrape cells from the surface of an agar plate with a coverslip. Fixation of samples directly from plates is a distinct advantage for the analysis of sexual differentiation as all stages of meiosis are present in one sample, thus obviating the needs for the analysis of multiple time points (16). It is not generally applicable, however, to mitotic growth, as the physiology of each cell is too variable.

2.4 Cell wall digestion

An important stage in the immunofluorescence microscopy of yeast cells is the removal of the cell wall. In most cases this is done after the cells have been fixed. It should be noted, however, that there are occasions when removal of the cell wall to make spheroplasts has been performed prior to fixing. This has been used for studies on SPB components in *S. cerevisiae* (12). In this case the cell wall was digested and the cells allowed to recover for about 30 min before solvent fixation.

There are several enzyme preparations available, of varying purities. The most common enzyme preparation used for cell wall digestion in both *S. pombe* and *S. cerevisiae* is Zymolyase 20T or 100T (ICN Biomedicals Inc.). It can be made up as a 10 mg/ml stock in phosphate buffer (pH 7.5) and used at a concentration of 0.4–1 mg/ml (see *Protocols 2* and *4* for detailed methods). This is a relatively pure preparation and is suitable for digestion of cell walls for yeast that have been growing in rich medium. Growth in minimal medium or certain mutations directly affect the composition of the cell wall such that digestion with this relatively pure enzyme preparation is not always sufficient. If growth in minimal medium cannot be avoided, due to the need, for example, to preserve selection of a plasmid, then two main approaches are possible. One is to grow the cells with selection until the final generation time. At this point the cells can be spun down, resuspended in rich medium and grown for a further 2–3 h. Alternatively, cruder enzyme preparations can be used. Such crude enzyme mixtures are likely to contain enzymes capable of digesting a wider range of sugar modifications and are likely, therefore, to result in a more complete removal of the cell wall. However, these preparations tend to be avoided by some researchers since they have been known to occasionally have significant levels of protease contamination which could be detrimental to later procedures.

2.5 Antibodies

2.5.1 Primary antibodies

There are an increasing number of antibodies available commercially. Some of these have been tried and tested for use in immunofluorescence or immunoelectron microscopy. Many others have only been shown to recognize proteins by Western blotting and in some cases by immunoprecipitation. It is important to note that success in one procedure by no means guarantees success for other purposes. Some companies do monitor use of their antibodies and their technical support can on occasion give useful advice on how to use antibodies for specific purposes or provide contact names for people who have bought and reported success for certain applications. In addition, certain websites or Internet newsgroups can be very useful sources to learn about recent technical advances and to find sources for specific antibodies (see *Table 1* for lists of websites). Commonly used, commercially available antibodies include those for tubulin (YOL1/34; Serotec), for nuclear pores (MAb414; BAbco) and for actin (N350; Amersham) although alternative antibodies which are not commercially available are often preferred.

In general it is still most common that antibodies to a protein of interest have not been raised. There are many excellent articles devoted to purification of proteins for the production of antibodies, including the Harlow and Lane manual (17) which has become a standard text in many labs. An increasing number of companies offer the facility of raising very high quality polyclonal sera at competitive prices. It is usually preferable to use the expertise of such companies if there are no experienced colleagues to supervise this sensitive and skilled procedure. For most immunofluorescence procedures purified polyclonal antibodies are as good as, or superior to, monoclonal antibodies. A preference for raising polyclonal antibodies derives from the probability that the serum will contain multiple antibody species that recognize different epitopes on the protein of interest. Thus, if a particular epitope is sensitive to fixation or deeply embedded in the native protein, it is also probable that other epitopes will not be. In addition, if an antiserum is raised to a large part of the protein it is possible that several antibodies could bind simultaneously, so increasing the strength of the immunofluorescence signal. When generating a polyclonal antiserum it is important to recognize that the serum is likely to contain many antibodies, some of which might react with other antigens in yeast cells. To ensure specificity, antibodies should be affinity purified. Micro-affinity purification of antibodies using antigen bound to nitrocellulose, and larger-scale, column-based techniques have been reported for the affinity purification of antibodies (33,17). Both approaches have been successfully used for producing antibodies for immunofluorescence in yeast cells. Another stage of purification that can be useful is to pre-adsorb the antiserum against fixed yeast cells in which the gene encoding the protein of interest has been deleted. If this is not possible then simply re-using an antibody can lead to a marked improvement on the second use as the non-specific background antibodies are adsorbed in the first application.

A more recent development in immunofluorescence microscopy has been the advent of numerous peptide tags which can be attached via recombinant DNA manipulations to the protein of interest (see section 3.1). Proteins

Table 2. Protein tags used for protein localization in yeast

Tag	Antibodies available	Supplier
Myc	9E10 mouse monoclonal (41)	Santa Cruz Biotechnology Inc.; hybridoma bank
	A14 rabbit polyclonal	Santa Cruz Biotechnology Inc.; hybridoma bank; BAbCo
HA	12CA5 mouse monoclonal	Santa Cruz Biotechnology Inc.; hybridoma bank;
	3F10 rat monoclonal[a]	Boehringer Mannheim
GST	B-14 mouse monoclonal or	Santa Cruz Biotechnology Inc.; hybridoma bank
	Z-5 rabbit polyclonal[b]	
His	MRGSHis[c]	Qiagen
	H-15 rabbit polyclonal	Santa Cruz Biotechnology Inc.; hybridoma bank
Protein-A	SPA-27 mouse monoclonal[d]	Sigma
GFP	mouse monoclonal or rabbit polyclonal	Clontech
Pk	336 (9)[e]	Serotec

[a] Low background for immunofluorescence.
[b] Alternatively, ask a colleague who has raised antibodies to a GST fusion protein for some sera and affinity purify with commercially available GST beads.
[c] This requires the MRGS sequence upstream of the 6xHis tag.
[d] Alternatively, use any rabbit immunoglobulin, followed by a fluorescent anti-rabbit antibody.
[e] To date this has been used primarily in fission yeast studies.

can then be visualized using readily available antibodies to these epitopes (*Table* 2).

The use of positive control antibodies is very valuable when attempting immunofluorescence for a protein of unknown cell localization. This will determine whether the technique has at least been successful for proteins which should be localized in a specific way. Particularly favoured benchmarks in both *S. cerevisiae* and *S. pombe* are the microtubule and actin cytoskeletons. Controlling the fixation of a particular preparation is of utmost importance as even standard aldehyde fixations do not always work routinely in expert immunofluorescence labs. Once a staining pattern has been observed it is imperative that the localization is then demonstrated to be specific and preferably seen when several different fixation procedures are used. In this regard, showing a lack of staining in a strain in which the gene encoding the protein of interest is deleted, is a suitable control. Another good control when using affinity-purified antiserum is to pre-adsorb the serum against the antigen and show that a specific staining pattern is no longer obtained. Green fluorescent protein (GFP) fusions now offer one of the best controls, as the distribution in living cells should offer relatively artefact-free results as long as expression levels are appropriate. Ultimately, there is no perfect control to show that the staining pattern reflects normal localization and it is important to confirm results by complementary approaches. These approaches might include co-immunoprecipitation to show binding to another component of the structure to which it is localized and co-fractionation *in vitro*.

2.5.2 Secondary antibodies

The choice of secondary antibody is critical for good staining. Two major considerations are the fluorophore of choice and the quality of the antibody preparations. The last 2–3 years have seen a plethora of new fluorophores available conjugated to a wide range of secondary antibodies. Advantages of some of the new fluorophores, such as Cy3 (Sigma) or the Alexa dyes (Molecular Probes; see *Table 1* for website), are: their brightness, their increased photostability and the fact that their excitation and emission spectra are sometimes much tighter and therefore less likely to 'bleed' into other channels, which is of particular importance for double-labelling experiments (see section 2.5.3). It should be remembered, however, that a brighter secondary is not going to overcome problems of a poor primary antibody. Rather, it is of particular use when the antigen is at low levels within the cells or if the primary antibody binds very specifically but weakly to its antigen. If the primary antibody is not sufficiently specific for the protein of interest, then a brighter secondary antibody will often result in an increase in general background staining but give no additional information. The increased background fluorescence might even mask an otherwise weak signal.

When considering purchasing secondary antibodies it is often a good idea to buy from several companies to allow evaluation of brightness, often balanced against an increased background of fluorescence. The level of cross-reactivity might also be important for particular experiments. As with primary antibodies, colleagues can often be the best source of advice and material. If a good source for relevant antibodies has been identified it is usually worth testing this for localizing your particular protein. Once tested, if suitable, then it is usually a good idea to note the batch number and order a number of vials (they can be stored at –80°C in 50% (v/v) glycerol for several years without any problems). Background fluorescence problems often arise because all animals suffer fungal infections and thus have antibodies that recognize yeast. The strength of this response varies from individual to individual. The problem is not always overcome by affinity purification to purified immunoglobulins because immunoglobulins offer such a range of epitopes. Many investigators therefore pre-adsorb their secondary antibodies with a crude protein extract before use. This can either be done when the antibody arrives prior to aliquoting for storage at –20°C or just before use. Any standard protein preparation technique that provides a concentrated protein sample is sufficient for this purpose (for examples see the yeast handbooks in the websites listed in *Table 1*).

2.5.3 Double labelling

To demonstrate the spatial relationships of two different proteins in the same cell, double labelling is often valuable. This can be relatively straightforward if primary antibodies from different types of animals are available. Appropriate secondary antibodies can then discriminate the antibody types to allow the staining patterns of each protein to be observed. It should be noted that the primary and secondary antibodies can be incubated with cells either sequentially or simultaneously. If the proteins show co-localization it is important to demonstrate that this is not due to cross-species reactivity of the secondary antibodies used. Thus, control samples should be set up in which one or the other primary antibody is left out of the incubation. Both secondary antibodies should then be added. A specific signal in the absence of the relevant primary antibody would indicate cross-reactivity. To remedy this, cross-reacting antibodies can be pre-absorbed against immobilized antibodies from the second class. A second concern with regard to double labelling is that of cross-over fluorescence. For example, sometimes illumination for fluorescein isothiocyanate (FITC) can result in some rhodamine fluorescence from a double-labelled sample. This can usually be circumvented by the use of different fluorophores with different excitation spectra further away from that of FITC (e.g. Cy5 or Texas Red) or by using different filter sets on the microscope. If it is necessary to use multiple antibodies from the same species to recognize different structures within the same cell, two approaches can be taken. If the signal from one of the primary antibodies is sufficiently strong, it may be possible to cut out the need for the amplification step with the secondary antibody and directly conjugate this primary antibody with a fluorochrome. The directly conjugated antibodies are then applied after processing for the other primary/secondary antibody combination. Alternatively one primary/ secondary combination is completed and then excess unlabelled secondary antibodies against the species of the primary antibody are used to block all remaining sites on this first primary antibody. The way is then clear for the next primary/secondary combination.

2.5.4 Epitope masking

One frequently encountered problem with protein localization is that the epitope recognized by the antibody is obscured by *in situ* interactions with other proteins. This can be circumvented by several approaches. A particularly useful step is a 5 min incubation with 0.2% SDS solution prior to addition of primary antibody (see *Protocol 3*). A second approach is treatment with a dilute trypsin solution to partially digest the sample. Finally it is possible to swell protoplasts with a mild osmotic shock and then fix in solvent to rapidly dehydrate the sample. This tends to rip open cell structures much more aggressively than standard fixation procedures.

2.6 Mounting cells

2.6.1 Immunofluorescence microscopy

There is a marked difference between the budding and fission yeast communities as to when cells are mounted. Each approach has its advantages and the reasons for the different approaches are largely historical. For most standard budding yeast procedures cells are generally mounted in wells on poly-L-lysine-coated slides after digestion of the cell wall. The remainder of the processing is done on the slide (see *Protocol 2*). For fission yeast, cells are not mounted until the staining is complete (*Protocol 4*). The advantage of the former procedure is that many samples can be easily processed by aspiration of the wash solution from the slide, and that relatively small volumes (10–15 μl) of antibodies are required for each sample. The advantage of latter approach is that many more cells are kept in reserve and if a particular mounting is upset (e.g. by moving the coverslip) the sample is not lost and it is trivial to remount another slide. In addition, glutaraldehyde/formaldehyde-fixed samples can be stored in solution for up to 5 years with little loss of staining if a strong fixation has been used.

For both budding and fission yeast a final step before the cells are to be visualized requires the addition of mounting medium. This is usually a glycerol-based solution that often contains a dye to permit the additional visualization of DNA (*Protocol 1*). We usually make our mounting solution in the lab using an anti-fade reagent, *p*-phenylenediamine. As this chemical is toxic and reputed to be carcinogenic, we prefer to weigh it out infrequently and make large batches which can be stored at –80 °C for long periods of time. In addition, there are various anti-fade mounting solutions commercially available, for example, Citifluor (Agar Scientific) and Vectashield (Vector Laboratories).

Protocol 1. Preparation of mounting medium for immunofluorescence procedures in yeast

Reagents

- *p*-Phenylenediamine (Sigma)
- Glycerol (BDH/Merck)
- 4′,6′-diamidino-2-phenylindole dihydrochloride (DAPI; Sigma), 1 mg/ml in water; store at –80 °C

- Phosphate-buffered saline (PBS): dissolve 8 g NaCl, 0.2 g KCl, 1.44 g Na_2HPO_4, 0.24 g KH_2PO_4 in 800 ml distilled H_2O. Adjust pH to 7.2 and make up to 1 litre. Autoclave.

Method

1. Add 100 mg *p*-phenylenediamine to 10 ml PBS. If the pH is below 9.0, bring it to pH 9.0 by adding 1m NaOH while stirring.

2. Add 90 ml glycerol and stir until homogeneous.

3. Add 2.25 μl of DAPI stock solution to allow nuclei to be visualized.

4. Store mounting medium at –80 °C in the dark. The solution should be discarded when it loses its clear colour and turns brown.

2.6.2 Cells expressing GFP-tagged proteins

The advent of GFP has seen a return of a number of traditional approaches for recording time-lapse images of yeast cells. One of the major concerns is the immobilization of yeast under physiologically relevant conditions. Many techniques are evolving but cells are generally either immobilized in 25% gelatin (18), in low melting temperature agarose (19), or on lectin-coated Petri dishes (20). The latter approach has the distinct advantage that it is possible to image the same cell during drug treatment and washout experiments (20). For further information on GFP imaging it would be most appropriate to consult one of the dedicated articles (18–20), as well as Chapters 5 and 6.

2.7 Visualizing cells

The small size of yeast cells means that the intensity of the signal is a major concern. It is important to have access to a good-quality microscope that has high-magnification objectives. The brighter the light source the better. Many fluorescence microscopes have a 50 W mercury lamp fitted as standard; however, if a 100 W light source is used there is a stunning difference in the image. The relationship between image intensity and the power of the light source is non-linear and the difference in quality is more than a simple doubling of intensity. Different objectives are designed for different purposes. Some objectives have extra lenses in order to generate a flat field in which the entire image is in focus. The more glass a lens has, the more light is lost, so it is preferable with dim samples to use lenses that have been optimized for light transmission rather than generating a flat field. Lenses that are not corrected are often not in favour with colleagues who work on larger cells and they may recommend against their use; however, because of small cell size of our samples such concerns are virtually irrelevant to the yeast cell biologist. The numerical aperture of a lens gives an indication of the amount of light transmitted by the lens: the higher the value, the more light is transmitted. The numerical aperture is usually written alongside the magnification power on the side of the objective. Some microscopes offer internal magnification in the microscope turret. The temptation is to use the greatest magnification possible. After a certain point, however, better images are generated by capture with a low magnification and increasing the image size in subsequent image- or film-processing steps.

2.8 Chromatin staining

The immunofluorescence protocols we cite here suggest DAPI for chromatin staining although propidium iodide or ethidium bromide can be used following RNase digestion. This becomes a particularily attractive option when is it is necessary to conduct double labelling with another stain that emits in the blue wavelengths such as Calcofluor or it is necessary to use a confocal microscope that does not have a UV laser.

3. Molecular manipulations

3.1 Using molecular tags to visualize proteins

The most significant recent impact on the study of protein localization in yeast has come from advances in protein tagging techniques. Tags are coding sequences that are fused to the gene encoding a protein of interest which either impart a useful property to the protein, such as the ability to associate with a partcicular matrix, or constitute epitopes recognized by specific monoclonal antibodies which are readily available. A range of tags are now available, from epitope tags such as Myc and HA, to larger tags such as glutathione S-transferase (GST) and fluorescent tags such as GFP. The tag to be used is attached via recombinant DNA manipulations to the protein of interest. Proteins can then be visualized, often using commercially available antibodies (see *Table 2*).

Proteins have been successfully localized in budding and fission yeast using a range of tags including the Myc tag (21,22) and the HA tag (23–25). An additional tag called Pk has recently been introduced in a series of tagging vectors for fission yeast, providing a third major working alternative (7–9,26). It should be noted that in most cases multiple copies of the tag need to be appended to the protein before the signal is sufficiently strong for immunofluorescence. In some cases as many as 18 copies have been added to a protein to enable detection in a particular assay (27). It should be noted that the HA tag can be adversely affected if using formaldehyde as a fixative, with signals being severely depleted after only 10 min fixation. In some cases the use of larger tags may facilitate one-step purification of proteins allowing parallel biochemical analyses to be carried out. Both GST and Protein A have been successfully used for protein localization (J. Pringle and S. Munro, personal communication). However, the large size of these tags may be a concern. A useful compromise between suitability for one-step purification and maintenance of protein function and localization in common use in fission yeast is to combine an epitope tag with multiple histidine residues to provide the potential for one-step purification on nickel resin (9,28). Two commercially available products which have come on to the market make His tagging a particularly attractive approach. These are the availability of 6xHis-tagged versions of virtually every gene in the budding yeast genome (Invitrogen; details are available on their website—see *Table 1*) and antibodies to multiple histidine residues (see *Table 2*).

There are very significant advantages to using tagging as an approach to localize proteins, including the relatively short time-scale needed to manipulate the gene and express it in cells compared with raising antibodies. In addition, provided that the antibody preparation is pure, only the protein carrying the tag will be detected by immunofluorescence. This is particularly relevant when studying one member of a family of highly related proteins.

Furthermore, being able to use the same antibodies for successive local-izations instead of having to raise antibodies for each protein of interest can be a considerable monetary saving.

The introduction of GFP is having a considerable impact upon the yeast community. It has allowed, for the first time, the localization of many proteins in real time. Furthermore, the persistence of GFP signal after fixation in 3% formaldehyde makes it a particularly attractive for several additional applica-tions. For example, it is possible to use it in mutant screens in which multiple living or fixed cells are directly imaged to identify mutants which have alterations in the distribution of a particular protein or identify genes based upon the localization of their GFP fusion proteins (29,30). It is possible, for example, to use a GFP-tagged protein to determine the conditions which preserve the *in vivo* distribution of a protein before embarking on the use of epitope tag or polyclonal antibodies. This offers a considerable time saving. Furthermore, anti-GFP polyclonal antibodies have been used by a number of groups in both immunofluorescence and immuno-electron microscopy.

Having made these points to extol the virtue of tagging as an approach, it is vital also to note the important factors that should be considered when using tagging to localize a protein.

3.2 Functionality

If a tag is to be used, it is important to demonstrate that the tagged protein is functional. This, however, is not always straightforward. If there is a change in phenotype caused by deletion or a mutation in the gene for the protein, then rescue of this phenotype is a good indication that the tag is not interfering with this function of the protein. It is essential to recognize though, that even if a tagged protein is capable of rescuing such a phenotype, it does not guarantee that all interactions are maintained, nor that they are as strong as in the wild-type situation. This caveat is probably the strongest argument against protein tagging and the one that should be most carefully addressed. As with many other techniques the use of complementary approaches can strenghten any data obtained. Using both antibodies and tagging approaches would lend support to data obtained from either individual approach. Furthermore, the insertion of the tag into different parts of the coding sequence will avoid some of the artefacts that come with blocking the function of a particular domain through appending a tag. Tagging at the N- or C-terminus is an option with most tagging vector sets, but transposon-mediated insertion of a tag offers a powerful tool to make tagging at different points within a protein a one-step, quick, easy and very advisable option (30–32). Some of these vectors which exploit the Cre–Lox site-specific recombinase system facilitate transposon flip-out, leaving the tag in-frame in the gene of interest (30). If it is not poss-ible to use both antibodies and tags to localize a particular protein it is important to demonstrate the preservation of other known characteristics, for example, demonstrating that a two-hybrid interaction can occur with a similar affinity for the tagged and untagged protein.

3.3 Expressing protein tags

Until relatively recently, most genes were manipulated to append the tag and then transformed into the yeast on a plasmid. Often the tagged proteins were over-expressed under the control of a variety of promoters. Such studies were open to criticism concerning the levels of the proteins being made and whether the localization was a true reflection of the physiological situation. In particular with GFP, the most brightly staining structures in a population are often the circular vacuoles that represent the breakdown of the over-expressed protein rather than the true localization of the molecule of interest. It is now possible and advisable to integrate many of the commonly used tags on to the genomic copy of the gene which means that it will be expressed under the control of its own promoter at wild-type levels. Advances in low-light-level detection of signals and in fluorophore technology have meant that even those proteins that are expressed at fairly low levels can often be detected, thus obviating the need to over-express proteins in order to localize them.

4. Localizing a novel protein in *S. cerevisiae*

As with many methods, each lab uses slight variations to optimize staining for particular antigens. It should be remembered, however, that artefactual staining is possible and varying the fixation procedures or the use of antibodies combined with GFP-fusion protein localization will give credance to any localization data. The first immunofluorescence method given here is the procedure that has been most widely used and is very similar to that described by Pringle *et al.* (14). It has been used for visualizing the actin cytoskeleton and several actin-binding proteins, for neck filament proteins, for proteins involved in secretion and for microtubules (e.g. ref. 33).

Protocol 2. Immunofluorescence in *S. cerevisiae* using formaldehyde fixation followed by a methanol–acetone treatment

Equipment and reagents

- Haemocytometer (Merck)
- Multiwell slides (ICN Biomedicals Inc.)
- Formaldehyde, 37% (w/v) solution (BDH/Merck)
- Potassium phosphate/sorbitol buffer: filter-sterilized 0.1 M potassium phosphate buffer pH 7.5 (BDH/Merck), with 1.2 M sorbitol (Sigma)

- Zymolyase 20T (ICN Biomedicals): make a 1 mg/ml stock in 0.1 M potassium phosphate buffer pH 7.5, and store at –20°C in 1 ml aliquots
- Poly-L-lysine hydrobromide, mol. wt 15 000–30 000 (Sigma P 7890); make up and store at –20°C as a 1 mg/ml stock in distilled H_2O

- β-Mercaptoethanol (Sigma)
- Methanol (BDH/Merck)
- Acteone (BDH/Merck)
- Blocking buffer: 1% bovine serum albumin (BSA), Fraction V (BDH/Merck) in PBS
- PBS (*Protocol 1*)
- Primary antibody
- Secondary antibody
- Mounting solution (*Protocol 1*)

Method

1. Grow a 5 ml culture of *S. cerevisiae* in rich medium to log phase (if you are counting cells using a haemocytometer, then this is a density of about 0.5–1 × 10^7 cells/ml which for wild-type cells corresponds to an OD$_{600nm}$ of about 0.2–0.5).

2. Fix the actively growing population of cells by adding 0.67 ml of 37% (w/v) formaldehyde and allow to stand at room temperature for at least 1 h.

3. Spin the cells for 2 min at 700g (about 2500 r.p.m. for a standard bench-top centrifuge) to pellet, and wash twice with 2.5 ml phosphate/sorbitol buffer.

4. Resuspend cells in 0.5 ml phosphate/sorbitol buffer and add 20 μl of 1 mg/ml Zymolyase stock and 1 μl β-mercaptoethanol. Incubate for 35 min at 37 °C.

5. During the cell wall digestion incubation time the slides can be prepared for cell mounting. Place 15 μl of 1 mg/ml poly-L-lysine on each well of a multi-well slide (pre-cleaned by immersing in distilled H$_2$O then in 95% (v/v) ethanol and air drying). Allow to sit for 5 min then wash each well three times with distilled water and air dry.

6. After cell wall digestion put 15 μl of cell suspension in each well. Allow to settle for 5–10 min then aspirate off gently. If obtaining enough cells for observation is a problem, the suspension can be spun down and resuspended in a smaller volume just before placing on the slide.

7. Place the slide in –20 °C methanol for 6 min, then in –20 °C acetone for 30 sec. This step flattens the cells and permeabilizes them.

8. After air drying the slides completely, wash each well ten times with blocking buffer by gently aspirating off the wash buffer and then re-applying from a pipette. From this step onward it is important not to let the cells become dry. Keep the slides in a humid environment, for example in a covered Petri dish next to a damp tissue.

9. Remove the final blocking buffer wash from the well and place 15 μl of the primary antibody on to the cells.[a] Incubate for at least 1 h.

10. Wash each well ten times with blocking buffer and add 15 μl of diluted secondary antibody.[b] Incubate for 1 h in a dark place to prevent bleaching of the fluorophore.

Protocol 2. *Continued*

11. Wash again ten times with blocking buffer, aspirate the buffer and put 5 µl of mounting solution into each well. Cover with coverslip, then seal around the edges with nail polish. Slides can be viewed immediately or stored at –20°C in the dark for several months.

---

[a] The antibody should be diluted as necessary in blocking buffer. In general antibodies are used at a greater concentration (about 10-fold greater) for immunofluorescence than for Western blotting.

[b] The secondary antibody should be diluted using blocking buffer. For many secondary antibodies, for example FITC- or rhodamine-conjugated goat anti-rabbit (Cappel/Organon Technika Inc.) we find that a 1/1000–1/2000 dilution is suitable. When using Cy3 (Sigma) we have most frequently used a 1/200 dilution.

While *Protocol 2* serves as a useful starting point, many antigens are better observed when this protocol is modified in some way. One of the most useful adaptations has been the replacement of the methanol–acetone step with a short incubation in a dilute SDS solution (*Protocol 3*). This has been particularly useful in *S. cerevisiae* for localizing many proteins which are found at the presumptive bud site, such as Cdc42p (34) and Sec4p (35). Other useful modifications might include shortening the formaldehyde fixation time (see section 2.3.2) or increasing the time of digestion of the cell wall or use a cruder enzyme preparation such as Novozym (Novo).

Protocol 3. Immunofluorescence in *S. cerevisiae* using formaldehyde fixation followed by an SDS treatment

Equipment and reagents
- See *Protocol 2*
- 0.2% SDS (Sigma)

Methods

1. Follow *Protocol 2*, steps 1–6.

2. Following mounting of the cells on the slide, aspirate excess cells and add 10 µl of 0.2% SDS in PBS for 3–5 min. Due to the presence of SDS the solution loses surface tension. To avoid spreading of solution from the wells, the volume of solution used in this permeabilization stage and in all subsequent wash steps and incubation procedures should be reduced to 10–12 µl rather than 15 µl.

3. Gently aspirate the SDS solution and wash ten times with blocking buffer. Do not leave the slide to dry in between the SDS step and washes.

4. Continue to process the cells for immunofluorescence as in *Protocol 2* from step 9.

---

5. How to localize a novel protein in *S. pombe*

Three standard fixation procedures are described. These are starting points that use an easily manageable number of cells, but it is possible to use one-tenth of the numbers stated with practice. Generally the fewer the cells in the antibody solution the better the images. Because all of these procedures involve so many centrifugation steps, it is important to ensure minimal cell loss by spinning each sample twice at all stages after the initial fixation step. On the first occasion the hinge of the microcentrifuge tube should face away from the axle whereas on the second spin it should face the axle.

Protocol 4. Combined glutaraldehyde–formaldehyde fixation

Equipment and reagents

- Haemocytometer (Merck)
- Fine forceps
- Blood tube rotator (SB1) (Stuart Scientific)
- Formaldehyde powder (BDH/Merck)
- Glutaraldehyde solution[a] (BDH/Merck)
- Sodium borohydride (BDH/Merck)
- 5 M NaOH
- PEM: 100 mM Pipes, 10 mM $MgSO_4$, 1 mM EGTA, adjusted to pH 6.9 with NaOH (all chemicals from BDH/Merck)
- PEMS: PEM + 1.2 M sorbitol (BDH/Merck); filter-sterilized

- PEMSZ: PEMS + 0.5 mg/ml Zymolyase 100-T (ICN Biomedicals)
- PEMST: PEMS + 1% (v/v) Triton X-100 (BDH/Merck)
- PEMBAL: PEM + 1% (w/v) BSA (essentially globulin free; Sigma), 0.1% (w/v) NaN_3 (BDH/Merck), 100 mM lysine–HCl (BDH/Merck)
- Mounting solution (*Protocol 1*) from which DAPI has been omitted
- PBS (*Protocol 1*)
- PBSD: PBS + 0.2 µg/ml DAPI (Sigma)
- PBSN: PBS + 0.1% (w/v) NaN_3 (BDH/Merck)

Method

1. Make up 10 ml of 30% formaldehyde.[b] To 3 g of paraformaldehyde add 10 ml PEM and heat in a water bath at 65°C for 5 min. Add 120 µl of 5 M NaOH and mix. Return to 65°C water bath until almost all of the formaldehyde has gone into solution.[c] Cool the solution to culture temperature by sitting on ice and centrifuge at 2500 r.p.m. in a bench-top centrifuge for 2 min to pellet any undissolved formal-dehyde.[d]

2. Fix the cells. Check the cell number of the culture and measure out sufficient culture into a 50 ml screw-capped plastic centrifuge tube to give a total of 7.5×10^7 cells. Add 0.125 vols of 30% (w/v) formal-dehyde solution to this fraction of the culture. Do this quickly and mix thoroughly. Thirty seconds to 1 min later, add glutaraldehyde to a final concentration of 0.2% (v/v),[e] mix well and agitate for 30 min.[f]

3. Digest the cell walls. Centrifuge the fixed culture preparation at 3000 r.p.m. in a bench-top centrifuge. Take up the pellet in 1 ml of PEM and transfer to a 1.5 ml microcentrifuge tube. Centrifuge the cells for 2 sec at 13000 r.p.m. in a microcentrifuge with the tube hinge

Protocol 4. *Continued*

facing away from the central axis. Turn the hinge around and pellet once more. Wash the pellet with PEM using this double spin each time and resuspend in 1.5 ml of PEMSZ to give a final concentration of 5×10^7 cells/ml. Incubate at 37 °C for 70 min.

4. *Permeabilization*: After 70 min in enzyme, pellet the cells and re-suspend in PEMST. After 30 sec, wash three times in PEM. Pellet and remove the solution.

5. *Quenching unreacted glutaraldehyde*: Add 5 ml of PEM to 5 mg[g] of sodium borohydride.[h] Resuspend the pellet in 200–300 μl of the 1 mg/ml sodium borohydride solution and leave *with the cap open* on the bench for 3–5 min.[i,j]

6. *Blocking any non-specific staining*: Wash the cells twice in PEM and pellet and resuspending the pellet in PEMBAL.[k] Place the cells on the blood tube rotator for 10 min.

7. *Antibody application*: Take one-tenth of the cell suspension, then pellet in a new 1.5 ml microcentrifuge tube, by centrifugation at 13000 r.p.m. Resuspend each pellet in 100 μl of antibody in PEMBAL.[l] Put the tubes on blood tube rotator and incubate overnight.

8. *Secondary antibody*: Pellet the cells. Remove the antibody solution and keep for subsequent re-use. Wash the cells twice in PEMBAL and resuspend in 100 μl of secondary antibody solution.[m] Incubate the tubes wrapped in aluminium foil at room temperature with constant rotation on blood tube rotator at room temperature or 37 °C for 5 h to overnight.

9. *DAPI staining*. Pellet cells from the secondary antibody and wash them once in PEMBAL. Wash once more in PBS and resuspend in PBSD. Pellet and resuspend in 40 μl of PBSN.[n,o]

10. Prepare large coverslips by washing them in detergent, water and finally acetone before drying them on filter paper.

11. Dry a mono-layer of cells on to a coverslip. Place the suspension on the coverslip and immediately remove as much of the suspension as possible by holding the coverslip at 45° to drain excess solution to one corner as you pipette the suspension off from that corner. Leave to dry.[p]

12. *Mounting*: Using fine forceps, invert the coverslip and hold it at a 45° angle over 5 μl of mounting medium on a clean microscope slide (wiping with ethanol is sufficient to clean the slide). Pull away the forceps so that the coverslip drops on to the mounting medium.[q,r]

13. *Imaging*: Check the fixation by examining the control sample.[s]

14. *Storage*: Cells that have been processed for combined aldehyde fixation can be stored in PBSN at 4°C for 4 years or more with only a marginal loss of signal intensity.

[a.] Check your stock, as the percentage varies between different batches and suppliers.

[b] Perform all stages of step 1 in a fume hood. Weigh out, store and dispose of formaldehyde solutions in a fume hood. **Caution**: wear gloves and eye protection when handling hot formaldehyde solutions.

[c] The solution should have a very light grey appearance.

[d] If the formaldehyde comes out of solution during the rest of the day, re-heat to 65°C.

[e] As with formaldehyde, dispense the glutaraldehyde in a fume hood, avoid breathing aldehyde fumes at all stages of the procedure and wear adequate eye protection and gloves.

[f] Fixation can be extended to 120 min, but 30 min is probably best.

[g] The static charges generated when using latex gloves with plastic tubes make it difficult to safely measure out sodium borohydride using a balance, so it is best to get a rough idea of how much 5 mg is on one occasion and then do it by eye on all subsequent occasions.

[h] Care should be taken at this step as the solution will bubble excessively when mixed with the sodium borohydride.

[i] It is important to leave the cap open because the pressure eventually forces the tops to open spontaneously, sending the sample everywhere.

[j] It is very important to spin the cells twice in sodium borohydride to avoid cell loss due to the bubbles in the solution.

[k] The lysine in PEMBAL reduces the background extensively and is one of the secrets of the success of this particular procedure (it also reduces non-specific interactions in Western blotting and immunoprecipitation).

[l] TAT1 (36) culture supernatant at a dilution of 1/20 serves as an excellent control antibody to check the fidelity of fixation.

[m] Generally either 1% (v/v) of a FITC-conjugated or 0.05–0.1% (v/v) of a Cy3-conjugated antibody.

[n] DAPI binds DNA much better in PBS than in PEMBAL, hence the wash in PBS prior to adding the DAPI.

[o] **Caution**: Take care when handling any solution containing sodium azide—it is as toxic as sodium cyanide.

[p] If in a hurry, dry high above a naked flame or with a hair dryer, but do not fry the cells! Heat damage will be seen as cells that have lost any specific fluorescence staining and have shrunk and shrivelled up to varying degrees.

[q] Do not move the coverslip once it has contacted the mounting medium, or else cells will detach from its surface and float around in free solution.

[r] If large coverslips have been used it is not necessary to seal with nail varnish as only the central region will be mounted. The thinner the layer of mounting medium the lower the level of background fluorescence.

[s] If TAT1 was used as a control, compare the staining with the images in Hagan and Yanagida (37). Microtubules extending from the outer face of the SPB at the end of anaphase indicate excellent fixation.

Common modifications to Protocol 4

Samples can be left in either primary or secondary antibody for 2 days at 4°C without affecting the final results; for some antigens this actually improves staining. The lysine in PEMBAL may be substituted with fish skin gelatin if it is known that the epitope contains multiple lysine residues.

Protocol 5. Formaldehyde fixation

Equipment and reagents

As for *Protocol 4*, without glutaraldehyde and sodium borohydride

Method

1. Make up formaldehyde according to *Protocol 4*, step 1.

2. Fix the cells. Check the cell number of the culture and measure out sufficient culture into a 50 ml screw-capped plastic centrifuge tube to give a total of 7.5×10^7 cells. Add 0.125 vols of 30% (w/v) formaldehyde solution to this fraction of the culture. Do this quickly and mix thoroughly. Agitate for 30 min.

3. Digest the cell walls. Centrifuge the fixed culture preparation at 3000 r.p.m. in a bench-top centrifuge. Take up the pellet in 1 ml of PEM and transfer to a 1.5 ml microcentrifuge tube. Centrifuge the cells at 13000 r.p.m. in a microcentrifuge with the tube hinge facing away from the central axis. Turn the hinge around and pellet once more. Wash the pellet with PEM using this double spin each time and resuspend in 0.75 ml of PEMS and add 0.75 ml PEMSZ. Incubate at 37°C for 70 min.

4. As for *Protocol 4*, step 4.

5. Proceed as from step 6 in *Protocol 4*.[a]

[a] Use the Amersham N350 anti-actin antibody as a control and compare your results with those reported by Marks and Hyams (38) to judge fixation. For best preservation of actin-related structures it is often best to substitute PM buffer for PEM; however, PEM is fine just for judging the quality of fixation by the presence of actin dots at the ends of the cell. The concentration of the primary and secondary antibodies is usually half of that required for combined aldehyde fixation.

Protocol 6. Solvent fixation

Equipment and reagents

- Fine forceps
- Buchner flask
- 30 ml glass sintered filter funnel
- 24 mm glass-fibre GFC filters (Whatman)
- Blood tube rotator (SB1) (Stuart Scientific)
- Methanol (BDH/Merck)
- PEM (see *Protocol 4*)
- PEMS (see *Protocol 4*)
- PEMSZ (see *Protocol 4*)
- PEMBAL (see *Protocol 4*)
- Mounting solution (*Protocol 1*) from which DAPI has been omitted
- PBSD (see *Protocol 4*)
- PBSN (see *Protocol 4*)

Method

1. Harvest cells from 30 ml of mid-log-phase culture by filtration on to glass-fibre filters (e.g. Whatman GFC series) or use a coverslip to scrape cells from agar plate.

2. Quickly plunge the filter pad/coverslip and cells into 5 ml cold methanol[a] in a 10 ml tube.

3. Place the tube in the freezer (−20 °C or −80 °C, the colder the better) for at least 1 h (overnight is also acceptable) then remove as much of the solvent as possible by careful aspiration and add PEM to a final volume of 5 ml. Spin and resuspend in 1 ml PEM.[b]

4. Transfer to a 1.5 ml microcentrifuge tube and pellet at 13000 r.p.m. (as for all subsequent washes). Resuspend in and wash twice more in PEM.

5. Resuspend in 0.75 ml PEMS. Add 0.75 ml PEMZ and incubate at 37 °C for 10 min.

6. Pellet and wash three times in PEM before resuspending in PEMBAL.

7. After 10 min in PEMBAL remove one-tenth of the suspension to a fresh 1.5 ml microcentrifuge tube. Spin the cells out of suspension in the fresh tube and resuspend in 100 μl PEMBAL containing antibody.[c,d]

8. After incubation of between 1 h and overnight, wash three times in PEMBAL and resuspend in 100 μl PEMBAL containing the secondary antibody.

9. After incubation between 1 h and overnight proceed as from *Protocol 4*, step 9.

[a] Methanol and tubes should be between −20 °C and −85 °C (the colder the better).
[b] If the cells are still stuck to the filter pad after 1 h in cold solvent, it may be possible to harvest them by lifting out the pad to a fresh microcentrifuge tube and washing the cells off with buffer.
[c] The concentration of both primary and secondary antibodies for solvent fixation is about one-fifth to one-tenth that required for combined aldehyde fixed samples (*Protocol 4*).
[d] Either TAT1 or N350 from Amersham (1/200 dilution) is a good control antibody for solvent fixation.

6. Localizing proteins and organelles by other fluorescence methods

Protocol 7. Rhodamine–phalloidin staining for filamentous actin in *S. cerevisiae*

Reagents

- Formaldehyde, 37% (w/v) solution (BDH/ Merck)
- Rhodamine–phalloidin (Molecular Probes): add methanol as directed by supplier and use immediately
- PBS + 1 mg/ml BSA (without Triton X-100)
- Multiwell slides (ICN Biomedicals Inc.)
- PBS containing 1 mg/ml BSA (Fraction V; BDH/Merck) + 0.1% (v/v) Triton X-100 (BDH/Merck)
- Poly-L-lysine hydrobromide, mol. wt 15000–30000 (Sigma P 7890)
- Mounting solution (*Protocol 1*)

Protocol 7. *Continued*

Method

1. Aliquot 1 ml of actively growing yeast culture in a microcentrifuge tube.

2. Add 134 μl of 37% (w/v) formaldehyde. Leave for 1 h.

3. Harvest the cells by spinning (2500 r.p.m.) and wash twice with PBS + 1 mg/ml BSA + 0.1% (v/v) Triton-X-100.

4. Take up pellet in 50 μl of PBS containing 1 mg/ml BSA + 0.1% (v/v) Triton-X-100 and add 20 μl of rhodamine–phalloidin.

5. Incubate for 30 min in the dark.

6. Wash twice with PBS + 1 mg/ml BSA (without Triton).

7. Place 15 μl of cell suspension on each well of a poly-L-lysine-coated slide (see *Protocol 2*) and allow the cells to settle for 5–10 min then aspirate off gently.

8. Wash the cells with PBS + 1 mg/ml BSA twice more then add a drop of mounting solution, cover with coverslip and seal.

Protocol 8. Rhodamine–phalloidin/Calcofluor staining to visualize F-actin and the cell wall in *S. pombe*

Equipment and reagents
- Fine forceps
- Blood tube rotator (SB1) (Stuart Scientific)
- 30% (w/v) formalin (Sigma)
- PM buffer: 35 mM KH_2PO_4, 0.5 mM $MgSO_4$, pH 6.8 (BDH/Merck)
- PMT: PM buffer + 1% (v/v) Triton X-100 (BDH/Merck)
- Calcofluor ('fluorescence brightener'; Sigma)
- Rhodamine–phalloidin (Molecular Probes)
- Mounting medium stock solution (*Protocol 1*)

Method

1. Fix 10 ml of mid-log-phase culture by adding 1.33 ml of 30% (w/v) formalin; agitate for 60 min.

2. Pellet cells by centrifugation and resuspend in 1 ml of PM buffer in a microcentrifuge tube.

3. Pellet cells by centrifugation at 13000 r.p.m. for 3 sec and wash in 1 ml of PM buffer. Repeat washing twice more and pellet the cells.

4. Resuspend pellet in PMT for 30 sec then pellet the cells again.

5. Wash twice in PM buffer and pellet.

6. Resuspend in 100 μl of 20 μg/ml rhodamine-conjugated phalloidin plus 2.5 mg/ml Calcofluor and place tubes, wrapped in aluminium foil, on a blood tube rotator for 1 h at room temperature, or overnight at 4°C.

7. Apply the cell suspension to a clean coverslip and remove as much as possible of the suspension to leave a monolayer of cells on the glass surface. After the cells have dried on to the surface, invert the coverslip over mounting medium made up from 90 μl of PM and 10 μl of frozen mounting stock solution (*Protocol 1*). It is important to substitute PM buffer for glycerol for the mounting medium for phalloidin staining.

Protocol 9. Using Lucifer Yellow for vital staining of vacuoles[a] (adapted from the protocol described by Dulic *et al.* (39))

Reagents

- 50 mM succinate/20 mM azide buffer (both from Sigma), brought to pH 5.0 with NaOH
- Lucifer Yellow, dilithium salt (Fluka): 40 mg/ml stock solution in water[b]

Method

1. Grow yeast cells to log phase (OD_{600nm} of about 0.2–0.5).

2. Take 1 ml of cells and spin, then resuspend the pellet in 30 μl of rich growth medium in a microcentrifuge tube.

3. Add 20 μl of 40 mg/ml Lucifer Yellow.

4. Shake at 30°C for 2 h.

5. Wash cells three times with 1 ml ice-cold succinate/azide buffer.

6. Resuspend in 10 μl of succinate/azide buffer and leave on ice until ready to be viewed.

7. To immobilize the cells it is sometimes helpful to add an equal volume of 1% molten low melting point agarose on a slide, then add the cells quickly and mount a coverslip.

[a] Vacuoles can also be labelled using the dye FM4-64 (Molecular Probes) following the method described by Vida and Emr (40).
[b] Store, wrapped in foil, at 4°C.

Protocol 10. DAPI staining and glutaraldehyde fixation to visualize chromatin

Equipment and reagents

- Glutaraldehyde solution (BDH/Merck)
- DAPI at 20 μg/ml (see *Protocol 1*)
- PBS (see *Protocol 1*)
- Aqueous *p*-phenylene-diamine mounting medium prepared as for *Protocol 1* except that PBS is substituted for glycerol in step 2

Protocol 10. *Continued*

Method

1. Fix 900 μl of cell culture in a 1.5 ml microcentrifuge tube by adding glutaraldehyde to a final concentration of 2.5%; place on ice for 5 min to 1 h.
2. Wash twice in 500 μl ice-cold PBS.
3. Resuspend in 100 μl PBS containing 20 μg/ml DAPI.
4. Dry on to clean coverslips as for *Protocol 4*, step 10, and mount on to aqueous *p*-phenylene-diamine mounting medium.

Protocol 11. Rhodamine-123 staining of mitochondria

Equipment and reagents

- Rhodamine 123 (Sigma), 1.5 mg/ml in methanol
- Aqueous mounting medium (*Protocol 10*)
- Blood tube rotator (SB1) (Stuart Scientific).

Method

1. To a 1 ml aliquot of culture in a 1.5 ml microcentrifuge add 10 μl of rhodamine 123 stock solution.
2. Incubate on blood tube rotator for 60 min.
3. Pellet the cells by centrifugation at 13000 r.p.m. for 10 sec.
4. Resuspend in 10–100 μl of culture medium depending upon culture density.
5. Air dry on to coverslip according to *Protocol 4*, step 10.
6. Invert on to 5 μl of aqueous mounting medium.

Protocol 12. Quick DAPI staining directly from plates

Equipment and reagents

- Toothpicks
- Yeast colonies
- Mounting solution (*Protocol 1*)
- Hot-plate stirrer

Method

1. Grow cells to produce a small colony on a plate and manipulate in the desired fashion, such as heat shift for a temperature-sensitive mutant.
2. Pick a colony into 5 μl of H$_2$O on a microscope slide.
3. Gently dry samples on to the slide using the hot plate of the heating block.
4. Place 5 μl of mounting medium on to the cells and cover with a coverslip.

Acknowledgements

We would like to thank members of both *S. cerevisiae* and *S. pombe* yeast communities for passing on useful tips on recent advances in the field. K.R.A. is supported by a Wellcome Trust Career Development Fellowship and I.M.H. by the Cancer Research Campaign.

References

1. Wigge, P. A., Jensen, O. N., Holmes, S., Soues, S., Mann, M. and Kilmartin, J. V. (1998). *J. Cell Biol.*, **141**, 967.
2. Ekwall, K. and Partridge, J. F. (1999). In *Chromosome Structural Analysis: A Practical Approach* (ed. W. A. Bickmore). Oxford University Press.
3. Adams, A. E. M. and Pringle, J. R. (1984). *J. Cell Biol.*, **98**, 934.
4. Hagan, I. M. and Hyams, J. S. (1988). *J. Cell Sci.*, **89**, 343.
5. Funabiki, H., Hagan, I., Uzawa, S. and Yanagida, M. (1993). *J. Cell Biol.*, **121**, 961.
6. Hagan, I., Hayles, J. and Nurse, P. (1988). *J. Cell Sci.*, **91**, 587.
7. Bridge, A. J., Morphew, M., Bartlett, R. and Hagan, I. M. (1998). *Genes Dev.*, **12**, 927.
8. Drummond, D. R. and Hagan, I. M. (1998). *J. Cell Sci.*, **111**, 853.
9. Craven, R. A., Griffiths, D. J. F., Sheldrick, K. S., Randal, R. E., Hagan, I. M. and Carr, A. M. (1998). *Gene*, **221**, 59.
10. Kilmartin, J. V., Dyos, S. L., Kershaw, D. and Finch, J. T. (1993). *J. Cell Biol.*, **123**, 1175.
11. Melan, M. A. and Sluder, G. (1992). *J. Cell Sci.*, **101**, 731.
12. Rout, M. P. and Kilmartin, J. V. (1990). *J. Cell Biol.*, **111**, 1913.
13. McDonald, K., Otoole, E. T., Mastronarde, D. N., Winey, M. and McIntosh, J. R. (1996). *Trends Cell Biol.*, **6**, 235.
14. Pringle, J. R., Adams, A. E. M., Drubin, D. G. and Haarer, B. K. (1991). In *Methods in Enzymology*, Vol. 194: *Guide to yeast genetics and molecular biology*, (eds C. Guthrie and G. R. Fink), p. 565. Academic Press, London.
15. Byers, B. (1981). In *Cytology of the Yeast Life Cycle* (ed. J. N. Strathern, E. W. Jones and J. R. Broach), p. 59. Cold Spring Harbor Laboratory Press, Cold Spring Harbor, NY.
16. Petersen, J., Nielsen, O., Egel, R. and Hagan, I. M. (1998). *J. Cell Sci.*, **111**, 867.
17. Harlow, E. and Lane, D. (1988). *Antibodies: A Laboratory Manual*. Cold Spring Harbor Laboratory Press, Cold Spring Harbor, NY.
18. Shaw, S. L., Yeh, E., Bloom, K. and Salmon, E. D. (1997). *Curr. Biol.*, **7**, 701.
19. Nabeshima, K., Saitoh, S. and Yanagida, M. (1997). In *Methods in Enzymology*, Vol. 283: *Cell cycle control* (ed. W. G. Dunphy), p. 459. Academic Press, London.
20. Ding, D.-Q., Chikashige, Y., Haraguchi, T. and Hiraoka, Y. (1998). *J. Cell Sci.*, **111**, 701.
21. Evan, G. I., Lewis, G. K., Ramsay, G. and Bishop, J. M. (1985). *Mol. Cell. Biol.*, **5**, 3610.
22. Terbush, D. R. and Novick, P. (1995). *J. Cell Biol.*, **130**, 299.

23. Bogerd, A. M., Hoffman, J. A., Amberg, D. C., Fink, G. R. and Davis, L. I. (1994). *J. Cell Biol.*, **127**, 319.

24. Funabiki, H., Yamano, H., Kumada, K., Nagao, K., Hunt, T. and Yanagida, M. (1996). *Nature*, **381**, 438.

25. Broughton, J., Swennen, D., Wilkinson, B. M., Joyet, P., Gaillardin, C. and Stirling, C. J. (1997). *J. Cell Sci.*, **110**, 2715.

26. Southern, J. A., Young, D. F., Heaney, F., Baumgartner, W. K. and Randall, R. E. (1991). *J. Gen. Virol.*, **72**, 1551.

27. Ciosk, R., Zachariae, W., Michaelis, C., Shevchenko, A., Mann, M. and Nasmyth, K. (1998). *Cell*, **93**, 1067.

28. Griffiths, D. J. F., Barbet, N. C., McCready, S., Lehmann, A. R. and Carr, A. M. (1995). *EMBO J.*, **14**, 5812.

29. Sawin, K. E. and Nurse, P. (1996). *Proc. Natl Acad. Sci. USA*, **93**, 15146.

30. Ross-Macdonald, P., Sheehan, A., Roeder, G. S. and Snyder, M. (1997). *Proc. Natl Acad. Sci. USA*, **94**, 190.

31. Morgan, B. A., Conlon, F. L., Manzanares, M., Millar, J. B. A., Kanuga, N., Sharpe, J., Krumlauf, R., Smith, J. C. and Sedgwick, S. G. (1996). *Proc. Natl Acad. Sci. USA*, **93**, 2801.

32. Ross-Macdonald, P., Sheehan, A., Friddle, C., Roeder, G. S. and Snyder, M. (1998). *Methods in Microbiol.*, **26**, 161.

33. Ayscough, K. R., Stryker, J., Pokala, N., Sanders, M., Crews, P. and Drubin, D. G. (1997). *J. Cell Biol.*, **137**, 399.

34. Ziman, M., Preuss, D., Mulholland, J., O'Brien, J. M., Botstein, D. and Johnson, D. I. (1993). *Mol. Biol. Cell*, **4**, 1307.

35. Brennwald, P. and Novick, P. (1993). *Nature*, **362**, 560.

36. Woods, A., Sherwin, T., Sasse, R., Macrae, T. H., Baines, A. J. and Gull, K. (1989). *J. Cell Sci.*, **93**, 491.

37. Hagan, I. and Yanagida, M. (1997). *J. Cell Sci.*, **110**, 1851.

38. Marks, J. and Hyams, J. S. (1985). *Eur. J. Cell. Biol.*, **39**, 27.

39. Dulic, V., Egerton, M., Elgundi, I., Raths, S., Singer, B. and Riezman, H. (1991). In *Methods in Enzymology*, Vol. 194: *Guide to yeast genetics and molecular biology* (eds C. Guthrie and G. R. Fink), p. 700. Academic Press, London.

40. Vida, T. A. and Emr, S. D. (1995). *J. Cell Biol.*, **128**, 779.

41. Evan, G. I., Lewis, G. K., Ramsey, G. and Bishop, J. M. (1985). *Mol. Cell. Biol.*, **5**, 3610.

Excitation and emission maxima of commonly used fluorophores

Fluorophore	Absorption maximum (nm)	Emission maximum (nm)	Observed colour
Alexa™ 350*	346	442	blue
Alexa™ 430*	433	539	yellow/green
Alexa™ 488*	495	519	green
Alexa™ 532*	531	554	yellow
Alexa™ 546*	556	575	orange
Alexa™ 568*	578	603	orange/red
Alexa™ 594*	590	617	red
Aminomethylcoumarin (AMCA)	345	445	blue
BODIPY™ FL	505	513	green
BODIPY™ TMR	542	574	red
BODIPY™ TR	589	617	red
Cascade Blue	400	420	blue
Cyanine (Cy2)	492	510	green
Indocarbocyanine (Cy3)	550	570	red
Indiodicarbocyanine (Cy5)	650	670	deep red
Fluorescein (FITC)	494	518	green
Hoechst 33258, 33342 (complexed to DNA)	352	461	blue
Lissamine Rhodamine B (LSRC)	570	590	red
Lucifer yellow	428	536	green
LysoTracker™ Green	504	511	green
LysoTracker™ Yellow	534	551	yellow
LysoTracker™ Red	577	592	red
MitoTracker™ Green FM	490	516	green
MitoTracker™ Orange CMTRos	551	576	orange
MitoTracker™ Red CMXRos	578	599	red
NBD	465	535	green
Oregon Green™ 488 fluorophore	496	524	green

Excitation and emission maxima of commonly used fluorophores

Fluorophore	Absorption maximum (nm)	Emission maximum (nm)	Observed colour
Oregon Green™ 500 fluorophore	503	522	green
Oregon Green™ 514 fluorophore	511	530	green
PKH2	490	504	green
PKH26	551	567	red
PKH67	490	502	green
Propidium iodide	536	617	red
R-Phycoerythrin	565	575	orange/red
Quantum Red™	488	670	red
Rhodamine (TRITC)	550	580	red
Rhodamine Red-X	570	590	red
Rhodol Green	499	525	green
Texas Red™	595	615	red

* Molecular Probes report that different anti-fade reagents and mounting media have variable effects on the fluorescence intensity and photostability of the Alexa™ range of fluorophores (data summarized at http://www.probes.com/lit/feature/alexa/section8.html). They do not list the effects of p-phenylenediamine, n-propylgallate or DABCO. The Alexa™ range has been specifically developed to take advantage of commonly used laser lines in confocal microscopy. A most useful database, giving spectra of a comprehensive list of fluorophores, together with those of filters and lasers, is available at http://www.fluorescence.bio-rad.com

A2

Commercial sources of antibodies

Where antibodies are listed as widely available, they can be purchased from several of the following: Affinity Bioreagents, BAbCO, Boehringer Mannheim, Chemicon, DAKO, Developmental Studies Hybridoma Bank, Santa Cruz Biotechnology, Serotec, Sigma, StressGen, TCS Biologicals, Transduction Laboratories, Zymed. This list of companies and antibodies is not intended to be exhaustive, but rather to provide a starting point for searching for suitable reagents. Some antibodies are species- or tissue-specific and may either not work at all on your particular sample, or may give an unexpected staining pattern, so caution should be used. The list of structures labelled has been taken mostly from suppliers' catalogues: only some have been tested by the editor for immunofluorescence studies.

1. Epitope tags

See Chapter 8, *Table 2*.

2. Cytoskeletal proteins

Antigen	Structure labelled	Company
Actin	actin filaments and monomer	widely available
Adhesion proteins	junctional structures, cell surface	widely available
Ankyrin	ankyrin	Zymed
Cytokeratins/keratins	keratin intermediate filaments	widely available
Cytoplasmic dynein intermediate chain	limited staining of spindle poles, spindle microtubules, vesicles	Chemicon
Dynactin p150[Glued]	limited staining of spindle poles, spindle microtubules, vesicles	Transduction Labs
Kinesin (mAb SUK4)	variable[a]	Developmental Studies Hybridoma Bank
Kinesin (mAbs H1, H2)	vesicles, intermediate compartment, Golgi apparatus	Chemicon

Antigen	Structure labelled	Company
Myosin	myosin	Boehringer, Chemicon, Developmental Studies Hybridoma Bank, Sigma
Neurofilaments	neurofilaments	widely available
Spectrin/fodrin	spectrin	Affinity Bioreagents, Boehringer, Chemicon
Tubulin	microtubules	widely available

[a] Golgi apparatus in some cells.

3. Organelles

Please note that some of these antibodies are very species-specific.

Antigen	Structure labelled	Company
p58	Golgi apparatus	Sigma
Adaptins (α, β, γ)	clathrin-coated membranes	Affinity Bioreagents, Sigma, Transduction Labs
ADP-ribosylation factor	Golgi apparatus	Affinity Bioreagents, Santa Cruz
Annexins	various	widely available
BiP	ER	Affinity Bioreagents
Calnexin	ER	Affinity Bioreagents, Chemicon, Santa Cruz, StressGen
Calreticulin	ER	Affinity Bioreagents, StressGen
Cathepsins	lysosomes	Chemicon, DAKO, Santa Cruz, Zymed
Clathrin heavy chain	clathrin-coated membranes	Affinity Bioreagents, Chemicon, Santa Cruz, Transduction Labs
βCOP	COP-I-coated membranes	Affinity Bioreagents, Sigma

Commercial sources of antibodies

Antigen	Structure labelled	Company
Dynamin	plasma membrane	Santa Cruz, StressGen, Transduction Labs.
EEA1	early endosomes	Santa Cruz
EGF receptor	plasma membrane, endosomes	Santa Cruz
Golgi glycoprotein	Golgi medial cisterna	Developmental Studies Hybridoma Bank
GS28 (GOS28)	Golgi apparatus	StressGen
HSP60 and mtHSP70	mitochondria	Affinity Bioreagents, Santa Cruz, Sigma, StressGen, Zymed
KDEL	ER	Affinity Bioreagents, StressGen
KDEL receptor	intermediate compartment	StressGen
LAMPs	lysosomes	Developmental Studies Hybridoma Bank, Santa Cruz, StressGen
Membrin (GS27)	Golgi apparatus	StressGen
Mitochondrial antigen	mitochondria	Chemicon
Nuclear lamins	nucleus	Developmental Studies Hybridoma Bank, Santa Cruz
Nuclear pore components	nuclear pores	Affinity Bioreagents
Protein disulfide isomerase	ER	Affinity Bioreagents, StressGen
Rabs	various organelles[a]	Santa Cruz, StressGen, Transduction Labs, Zymed
Rough ER glycoprotein	RER	Developmental Studies Hybridoma Bank
Synaptic vesicle components	synaptic vesicles	widely available
Syntaxin		Transduction Labs
TGN38 and 41	*trans*-Golgi network	Affinity Bioreagents
Transferrin receptor	endocytic pathway	DAKO, Serotec, Zymed

[a] Each Rab is specific for a particular organelle.

A3

Optical units

To a physicist, wishing to discuss resolution, the final magnification of an image is unimportant. In theoretical papers on microscopy, features in an image are often measured in units which scale with the Airy disc size. It would be possible to use the Airy disc diameter as such a unit in the x–y plane, but it is more customary to use a 'radial optical unit' (o.u.), defined by

$$\text{o.u.} = \lambda M_{\text{tot}} / 2\pi \, \text{NA}$$

where λ is the wavelength of the light, M_{tot} is the total magnification of the image (objective magnification times any other magnification factor in the system) and NA is the numerical aperture.

A4

List of suppliers

Affinity Bioreagents Inc.
14818 West 6th Avenue, Suite 10A, Golden, CO 80401, USA. Tel.: +1-303-2784535; fax: +1-303-278-2424; website: http://www.bioreagents.com/affinity

Agar Scientific
66A Cambridge Road, Stansted, Essex CM26 8DA, UK. Tel.: +44-(0)1279-813519; fax: +44-(0)1279-815106.

Air Products and Chemicals, Inc.
7201 Hamilton Boulevard, Allentown, PA 18195-1501, USA. Tel.: +1-610-481-4911; fax: +1-610-481-5900; website: http://www.airproducts.com/

Amersham
Amersham International plc., Lincoln Place, Green End, Aylesbury, Buckinghamshire HP20 2TP, UK.
Amersham Corporation, 2636 South Clearbrook Drive, Arlington Heights, IL 60005, USA.

Anderman
Anderman and Co. Ltd., 145 London Road, Kingston-Upon-Thames, Surrey KT17 7NH, UK.

Applied Precision Inc.
1040 12th Avenue NW, Issaquah, WA 98027, USA. Tel.: +1-425-557-1000; fax: +1-425-557-1055; e-mail: info@api.com; website: http://www.api.com/dv.html

BAbCo (Berkeley Antibody Company)
4131 Lakeside Drive, Richmond, CA 94806, USA. Tel.: +1-510-222-4940; fax: +1-510-222-1867.

Beckman Instruments
Beckman Instruments UK Ltd., Oakley Court, Kingsmead Business Park, London Road, High Wycombe, Bucks HP11 1J4, UK.
Beckman Instruments Inc., PO Box 3100, 2500 Harbor Boulevard, Fullerton, CA 92634, USA.

Becton Dickinson
Becton Dickinson and Co., Between Towns Road, Cowley, Oxford OX4 3LY, UK.

Becton Dickinson and Co., 2 Bridgewater Lane, Lincoln Park, NJ 07035, USA.

Bio

Bio 101 Inc., c/o Statech Scientific Ltd, 61–63 Dudley Street, Luton, Bedfordshire LU2 0HP, UK.

Bio 101 Inc., PO Box 2284, La Jolla, CA 92038–2284, USA.

Biological Optical Technologies (Bioptec Inc.)

3560 Beck Road, Butler, PA 16002, USA. Tel.: +1-724-282-7145; fax: +1-724-282-0745; e-mail: info@bioptechs.com; website: http://www.bioptechs.com/

Bio-Rad

Bio-Rad Laboratories Ltd: Bio-Rad House, Maylands Avenue, Hemel Hempstead, Hertfordshire HP2 7TD, UK. Tel.: +44-(0)181-328-2000; FreePhone: 0800-181134; fax: +44-(0)1442-259118; website: http://www.biorad.com

Life Science Research: 2000 Alfred Nobel Drive, Hercules, CA 94547, USA. Tel.: +1-510-741-1000; fax: +1-510-741-5800;

Bitplane AG

Technoparkstrasse 1 CH-8005, Zürich, Switzerland. Tel.: +41-1-440-29-65; fax: +41-1-445-15-41; website: http://www.bitplane.ch

Boehringer Mannheim

Boehringer Mannheim UK (Diagnostics and Biochemicals) Ltd, Bell Lane, Lewes, East Sussex BN17 1LG, UK.

Boehringer Mannheim Corporation, Biochemical Products, 9115 Hague Road, P.O. Box 504 Indianapolis, IN 46250–0414, USA.

Boehringer Mannheim Biochemica, GmbH, Sandhofer Str. 116, Postfach 310120 D-6800 Ma 31, Germany.

British Drug Houses (BDH) Ltd, Poole, Dorset, UK.

Calbiochem-Novabiochem (UK) Ltd

Boulevard Industrial Park, Padge Road, Beeston, Nottingham NG9 2JR, UK. Tel.: +44-(0)800-622-935 or +44-(0)115-943-0840; fax: +44-(0)115-943-0951; website: http://www.cnuk.co.uk

Cappell/Organon Technika Inc.

100 Akzo Avenue, Durham, NC 27712, USA. Tel.: +1-919-620-2000; fax: +1-919-620-2107.

Carl Zeiss Ltd

PO Box 78, Woodfield Road, Welwyn Garden City, Herts AL7 1LU, UK. Tel.: +44-(0)1707-871200; fax: +44-(0)1707-871287; website: http://www.zeiss.co.uk

Tatzenpromenade 1a, D-07745 Jena, Germany. Tel.: +49-(0)364164-2873; fax.: +49-(0)364164-3192.

(R.P.) Cargille Labs Inc.

55 Commerce Road, Cedar Grove, NJ 07009-1289, USA. fax: +1-201-239-6096.

Chemicon International

Chemicon International Ltd: 2 Admiral House, Cardinal Way, Harrow HA3 5UT, UK. Tel.: +44-(0)181-8630415; fax: +44-(0)181-8630416.

28835 Single Oak Drive, Temecula, CA 92590, USA. Tel.: +1-909-676-8080; fax: +1-909-676-9209; website: http://www.chemicon.com

Chroma Technology Corp.

72 Cotton Mill Hill, Unit A-9, Brattleboro, VT, USA. Tel.: +1-802-257-1800; fax: +1-802-257-9400; e-mail: sales@chroma.com;website:http://www.chroma.com

Clontech

Clontech Laboratories Inc.: 1020 East Meadow Circle, Palo Alto, CA 94303-4230, USA. Tel: +1-415-424-8222; fax: +1-415-424-1064; website: http://www.clontech.com

Clontech Laboratories UK Ltd: Unit 2, Intec 2, Wade Road, Basingstoke, Hants RG24 8NE, UK. Tel.: +44-(0)1256-476500; fax: +44-(0)1256-476499.

DAKO Ltd

16 Manor Courtyard, Hughenden Avenue, High Wycombe, Bucks HP13 5RE, UK. Tel.: +44-(0)1494-452106; fax: +44-(0)1494-441846; website: http://www.dako.com/

Developmental Studies Hybridoma Bank

Department of Biological Sciences, The University of Iowa, 436 Biology Building, Iowa City, IA 52242, USA. Tel.: +1-319-335-3826; fax: +1-319-335-2077; website: http://www.uiowa.edu/~dshbwww/

Difco Laboratories

Difco Laboratories Ltd., P.O. Box 14B, Central Avenue, West Molesey, Surrey KT8 2SE, UK.

Difco Laboratories, P.O. Box 331058, Detroit, MI 48232–7058, USA.

Du Pont

Dupont (UK) Ltd., Industrial Products Division, Wedgwood Way, Stevenage, Herts, SG1 4Q, UK.

Du Pont Co. (Biotechnology Systems Division), P.O. Box 80024, Wilmington, DE 19880–002, USA.

EM Sciences

Science Services Ltd, Greenwood House 4/7, Salisbury Court, London EC4Y 8BT, UK. Tel.: +44-(0)171-7369927; fax: +44-(0)171-7369974.

Eppendorf

Eppendorf-Netheler-Hinz GmbH: 22331 Hamburg, Germany. Tel.: +49-(0)40-53801-0; fax: +49-(0)40-53801556; website: http://www.eppendorf.com/

UK office: 10 Signet Court, Swanns Road, Cambridge, CB5 8LA, UK; Tel.: +44-(0)1223-722100; fax: +44-(0)1223-722002.

European Collection of Animal Cell Culture, Division of Biologics, PHLS Centre for Applied Microbiology and Research, Porton Down, Salisbury, Wilts SP4 0JG, UK.

Falcon (Falcon is a registered trademark of Becton Dickinson and Co.).

Fisher Scientific Co., 711 Forbest Avenue, Pittsburgh, PA 15219–4785, USA.

Flow Laboratories, Woodcock Hill, Harefield Road, Rickmansworth, Herts. WD3 1PQ, UK.

Fluka

Fluka-Chemie AG, CH-9470, Buchs, Switzerland.

Fluka Chemicals Ltd., The Old Brickyard, New Road, Gillingham, Dorset SP8 4JL, UK.

Gibco BRL

Gibco BRL (Life Technologies Ltd.), Trident House, Renfrew Road, Paisley PA3 4EF, UK.

Gibco BRL (Life Technologies Inc.), 3175 Staler Road, Grand Island, NY 14072–0068, USA.

Halocarbon Products Corp.

887 Kinderkamack Road, River Edge, NJ 07661, USA. Tel.: +1-201-262-8899; fax: +1-201-262-0019.

European supplier: KMZ Chemical Ltd, 48 Station Road, Stoke D'Abernon, Cobham, Surrey KT11 3BN, UK. Tel.: +44-(0)1932-866426; fax: +44-(0)1932-867099.

Hamamatsu Photonics UK Ltd

Lough Point, 2 Gladbeck Way, Windmill Hill, Enfield, Middlesex EN2 7JA, UK. Tel.: +44-(0)181-3673560; fax: +44-(0)181-3676384.

Arnold R. Horwell, 73 Maygrove Road, West Hampstead, London NW6 2BP, UK.

Hybaid

Hybaid Ltd., 111–113 Waldegrave Road, Teddington, Middlesex TW11 8LL, UK.

Hybaid, National Labnet Corporation, P.O. Box 841, Woodbridge, NJ. 07095, USA.

HyClone Laboratories 1725 South HyClone Road, Logan, UT 84321, USA.

ICN Biomedicals Ltd

Unit 18, Thame Park Business Centre, Wenman Road, Thame, Oxon OX9 3XA, UK.

ICN Pharaceuticals

Biomedical Research Products, 3300 Hyland Avenue, Costa Mesa, CA 92626, USA. Tel: +1-714-545-0100; fax: +1-714-557-4872; website: http://www.icnpharm.com/

InerFocus Ltd (Fine Surgical Tools)

14/15 Spring Rise, Falconer Road, Haverhill, Suffolk CB9 7XU, UK. Tel: +44-(0)1440-703460; fax: +44-(0)1440-704397.

Inovision Corp.

2810 Meridian Parkway, Suite 148, Durham, NC 27713, USA. Tel.: +1-919-361-4609; fax: +1-919-361-5876; website: http://www.inovis.com

Intelligent Imaging Innovations

1620 Market Street, Suite 1E, Denver, CO 80202, USA. Tel +1-303-607-9429; fax: +1-303-607-9430; website: http://www.intelligent-imaging.com/home.html

International Biotechnologies Inc., 25 Science Park, New Haven, Connecticut 06535, USA.

Intracel Ltd
Unit 4, Station Road, Shepreth, Royston, Herts SG8 6PZ, UK. Tel.: +44-(0)1763-262680; fax: +44-(0)1763-262676.

Invitrogen Corporation
Invitrogen Corporation 3985 B Sorrenton Valley Building, San Diego, CA. 92121, USA.
Invitrogen Corporation c/o British Biotechnology Products Ltd., 4–10 The Quadrant, Barton Lane, Abingdon, OX14 3YS, UK.

Jackson ImmunoResearch Laboratories, Inc.
872 West Baltimore Pike, PO Box 9, West Grove, PA 19390, USA. Tel.: +1-610-869-4024; fax: +1-610-869-0171; website: http://www.jacksonimmuno.com/index.htm
UK distributor: Stratech Scientific Ltd, 61–63 Dudley Street, Luton, Beds LU2 0NP, UK. Tel.: +44-(0)1582-481884; fax: +44-(0)1582-481895.

Kinetic Imaging Ltd
South Harrington Building, Sefton Street, Liverpool L3 4BQ, UK; Tel.: +44-(0)151-7098661; fax: +44-(0)151-7098633; website: http://www.kineticimaging.com/contact.htm

Kodak: Eastman Fine Chemicals 343 State Street, Rochester, NY, USA.

Laboratory Sales (UK) Ltd (LSL)
Unit 20–21, Transpennine Trading Estate, Rochdale OL11 2PX, UK. Tel.: +44-(0)706-356444; fax: +44-(0)706-860885.

Leica Microsystems (UK) Ltd
Davy Avenue, Knowhill, Milton Keynes MK5 8LB, UK. Tel.: +44-(0)1908-246246; fax: +44-(0)1908-609992; website: http://www.leica.co.uk/

Ludl Electronic Products Ltd
171 Brady Avenue, Hawthorne, NY 10532, USA. Tel.: +1-914-769-6111; fax: +1-914-769-4759; website: http://www.ludl.com

Life Technologies Inc., 8451 Helgerman Court, Gaithersburg, MN 20877, USA.

MatTek Corporation
200 Homer Avenue, Ashland, MA 01721, USA. Tel.: +1-508-881-6771; fax: +1-508-879-1532; website: http://www.mattek.com/

Merck
Merck Industries Inc., 5 Skyline Drive, Nawthorne, NY 10532, USA.
Merck, Frankfurter Strasse, 250, Postfach 4119, D-64293, Germany.

Millipore
Millipore (UK) Ltd., The Boulevard, Blackmoor Lane, Watford, Herts WD1 8YW, UK.
Millipore Corp./Biosearch, P.O. Box 255, 80 Ashby Road, Bedford, MA 01730, USA.

Molecular Probes

Molecular Probes Inc. (USA): PO Box 22010, 4849 Pitchford Avenue, Eugene, OR 97402-9165, USA. Tel: +1-541-465-8300; fax: +1-541-344 6504; e-mail: order@probes.com; website: http://www.probes.com; for fluorescent beads see http://www.probes.com/handbook/ch26-1.html

Molecular Probes Europe BV: PoortGebouw, Rijnsburgerweg 10, 2333 AA Leiden, The Netherlands. Tel: +31-71-523-3378; fax: +31-71-523-3419; e-mail: eurorder@probes.nl

New England Biolabs (NBL)

New England Biolabs (NBL), 32 Tozer Road, Beverley, MA 01915–5510, USA.

New England Biolabs (NBL), c/o CP Labs Ltd., P.O. Box 22, Bishops Stortford, Herts CM23 3DH, UK.

Nikon Corporation, Fuji Building, 2–3 Marunouchi 3-chome, Chiyoda-ku, Tokyo, Japan.

Olympus Optical Co. (UK) Ltd

Microscope Division, Great Western Industrial Park, Dean Way, Southall, Middlesex UB2 4SB, UK. Tel.: +44-(0)171-2500179; fax: +44-(0)171250 4677; website: http://www.olympus.com

Omega Optical Inc.

3 Grove Street, Brattleboro, VT 05301, USA. Tel.: +1-802-254-2690; fax: +1-802-254-3937; website: http://www.omegafilters.com/

UK distributor: Glen Spectra Ltd, 2–4 Wigton Gardens, Stanmore, Middlesex HA7 1BG, UK. Tel: +44-(0)181-2049517; fax: +44-(0)181-2045189; website: http://www.isa-gs.co.uk

Oxyrase Inc.

PO Box 1345, Mansfield, OH 44901, USA. Tel.: +1-419-589-8800; fax: +1-419-589-9919; website: http://www.oxyrase.com

PCO Computer Optics GmbH

Ludwigsplatz 4, 93309 Kelheim, Germany. Tel.: +49-9441-200-50; fax: +49-9441-200-520; website: http://www.pco.de/index.html

Perkin-Elmer

Perkin-Elmer Ltd., Post Office Lane, Beaconsfield, Bucks, HP9 1QA, UK.

Perkin-Elmer-Cetus (The Perkin-Elmer Corporation), 761 Main Avenue, Norwalk, CT 0689, USA.

Pharmacia Biotech Europe Procordia EuroCentre, Rue de la Fuse-e 62, B-1130 Brussels, Belgium.

Pharmacia Biosystems

Pharmacia Biosystems Ltd. (Biotechnology Division), Davy Avenue, Knowlhill, Milton Keynes MK5 8PH, UK.

Pharmacia LKB Biotechnology AB, Björngatan 30, S-75182 Uppsala, Sweden.

Photonic Science

Millham, Mountfield, Robertsbridge, East Sussex TN32 5LA, UK. Tel.: +44-

(0)1580-881199; fax: +44-(0)1580-880-910; website: http://www.photonic-science.ltd.uk

Pierce & Warriner (UK) Ltd
44 Upper Northgate Street, Chester CH1 4EF, UK. Tel.: +44-(0)1244-382525; fax: +44-(0)1244-373212; website: http://www.piercenet.com

Promega
Promega Ltd., Delta House, Enterprise Road, Chilworth Research Centre, Southampton, UK.
Promega Corporation, 2800 Woods Hollow Road, Madison, WI 53711–5399, USA.

Qiagen
Qiagen Inc., c/o Hybaid, 111–113 Waldegrave Road, Teddington, Middlesex, TW11 8LL, UK.
Qiagen Inc., 9259 Eton Avenue, Chatsworth, CA 91311, USA.

Roche Diagnostics (*see* Boehringer Mannheim)

Roper Scientific (Photometrics and Princeton Instruments)
UK: PO Box 1192, 43 High Street, Marlow, Bucks SL7 1GB, UK. Tel.: +44-(0)1628-890858; fax: +44-(0)1628-898381.
USA: 3660 Quakerbridge Road, Trenton, NJ 08619, USA. Tel.: +1-609-587-9797; fax: +1-609-587-1970; website: http://www.photomet.com/cameras/ or http://www.prinst.com

Santa Cruz Biotechnology Inc.
2161 Delaware Avenue, Santa Cruz, CA 95060, USA. Tel: +1-831-457-3800; fax: +1-831-457-3801; website: http://www.scbt.com/

Scanalytics Inc.
8550 Lee Highway, Suite 400, Fairfax, VA 22031-1515, USA. Tel: +1-703-208-2230; fax: +1-703-208-1960; e-mail: info@scanalytics.com; website: http://www.scanalytics.com

Schleicher and Schuell
Schleicher and Schuell Inc., Keene, NH 03431A, USA.
Schleicher and Schuell Inc., D-3354 Dassel, Germany. Schleicher and Schuell Inc., c/o Andermann and Company Ltd.

Serotec
Serotec Ltd: 22 Bankside, Station Approach, Kidlington, Oxford OX5 1JE, UK. Tel +44-(0)1865-852722; fax: +44-(0)1865-373899; e-mail: serotec@serotec.demon.co.uk
Serotec Inc.: Partners 1, 1017 Main Campus Drive, Suite 2450, NCSU, Raleigh, NC 27606, USA. Tel: +1-800-265-7376; fax: +1-919-515-9980; e-mail: serotec@serotec-inc.com

Shandon Scientific Ltd., Chadwick Road, Astmoor, Runcorn, Cheshire WA7 1PR, UK.

Sigma Chemical Company
Sigma Chemical Company (UK), Fancy Road, Poole, Dorset BH17 7NH, UK.

Sigma Chemical Company, 3050 Spruce Street, P.O. Box 14508, St. Louis, MO 63178–9916.

Sorvall DuPont Company, Biotechnology Division, P.O. Box 80022, Wilmington, DE 19880–0022, USA.

Stratagene

Stratagene Ltd., Unit 140, Cambridge Innovation Centre, Milton Road, Cambridge CB4 4FG, UK.

Strategene Inc., 11011 North Torrey Pines Road, La Jolla, CA 92037, USA.

StressGen Biotechnologies Corp.

120-4243 Glanford Avenue, Victoria, BC, Canada V87 4B9. Tel: +1-250-744-2811; fax: +1-250-744-2877; website: http://www.stressgen.com/reagents

UK suppliers: Bioquote Ltd, The Raylor Centre, James Street, York YO10 3DW, UK.

Stuart Scientific

Holmethorpe Avenue, Holmethorpe Industrial Estate, Redhill, Surrey RH1 2NB, UK.

TAAB Laboratories Equipment Ltd

3 Minerva House, Calleva Park, Aldermaston, Berks RG7 8NA, UK. Tel: +44-(0)118-9817775; fax: +44-(0)118-9817881; website: http://www. microscopy-uk.org.uk/prodir/taab.html

Technical Video Ltd

PO Box 693, Woods Hole, MA 02543, USA. Tel: +1-508-563-6377; fax: +1-508-563-6265; e-mail: rknudson@tiac.net; website: http://www. technicalvideo.com

Transduction Laboratories Inc.

133 Venture Ct, Suite 5, Lexington, KY 40511-2624, USA. Tel.: +1-606-259-1550; fax: +1-606-259-1413; website: http://www.translab.com

UK distributors: Affiniti Research Products Ltd, Mamhead Castle, Mamhead, Exeter EX6 8HD, UK. Tel.: +44-(0)1626-891010; fax: +44-(0)1626-891090; website: http://www.affiniti-res.com

United States Biochemical, P.O. Box 22400, Cleveland, OH 44122, USA.

VayTek Inc.

305 West Lowe Avenue, Suite 109, Fairfield, IA 52556, USA. Tel +1-515-472-2227; fax: +1-515-472-8131; e-mail vaytek@vaytek.com; website: http:// www.vaytek.com/

Vector Laboratories

USA: 30 Ingold Road, Burlingame, CA 94010, USA. Tel: +1-650-697-3600; fax: +1-650-697-0339; e-mail: vector@vectorlabs.com; website: http://www. vectorlabs.com/

Vector Laboratories UK: 16 Wulfric Square, Bretton, Peterborough PE3 8RF, UK. Tel: +44-(0)1733-265530; fax: +44-(0)1733-263048; e-mail: vector@ vectorlabs.co.uk

Wellcome Reagents, Langley Court, Beckenham, Kent BR3 3BS, UK.

Whatman International Ltd

20 St Leonard's Road, Maidstone, Kent ME16 0LS, UK. Tel: +44-(0)1622-674821; fax: +44-(0)1622-682288.

Yellow Springs Instuments (YSI)

Yellow Springs Instuments Co. Inc.: Yellow Springs, OH 45387, USA. Tel: +1-513-767-7241.

YSI Ltd (UK): Lynchford House, Lynchford Lane, Farnborough, Hampshire GU14 6LT, UK. Tel: +44-(0)1252 514711; fax: +44-(0)1252 511855.

Zymed Laboratories Inc.

458 Carlton Court, South San Francisco, CA 94080, USA. Tel.: +1-650-871-4494; fax: +1-650-871-4499. Website: www.zymed.com

UK distributors: Cambridge Bioscience, 24–25 Signet Court, Newmarket Road, Cambridge CB5 8LA, UK. Tel.: +44-(0)1223-316855; +44-(0)1223-360732.

Other useful websites

Adobe: http://www.adobe.co.uk/
Apple Macintosh: http://www.apple.com/
Epson: http://www.epson.com/

For microscopy:

http://mc11.mcri.ac.uk/microscopy.html
http://photonics.usc.edu/bobc/yellowpgs/photonics.html
http://www.videomicroscopy.com/
http://www3.bc.sympatico.ca/micron/microscope/
http://www.biotech.ufl.edu/%7Eemcl/hotlinks.html
http://www.ou.edu/research/electron/www-vl/
http://www.cs.ubc.ca/spider/ladic/confocal.html
http://www.bocklabs.wisc.edu/imr/home.htm

Glossary of microscopy terms

acousto-optical deflector (AOD) a solid-state device by which a light beam may be scanned under electronic control.

avalanche photodiode a p–n photodiode operated with a reverse bias of approximately 100 V, giving a large intrinsic amplification of photo-current.

charge-coupled device (CCD) a form of electronic camera in which the light is detected by an array of MOS junctions or photodiodes.

C-mount a standard threaded mounting for camera lenses (dimensions are given in the text).

continuous wave (CW) this refers to lasers in which emission is continuous, not pulsed.

dichroic a coated mirror that is angled at $45°$ to the light path, and which reflects light of wavelengths shorter than the specified cut-off, and transmits longer wavelengths.

full-width half-maximum (FWHM) a simple measure of spatial or spectral resolution, being the width of a peak at half its maximum height above the baseline.

hole-accumulation diode (HAD) a pinned photodiode with properties of both a gate and a photodiode. Hyper-HAD, a Sony trademark, refers to a CCD chip which lies at the basis of many commercially produced cameras in which microlenses are used to increase the fill factor of an HAD array.

intensified charge-coupled device (ICCD) a camera which combines an image intensifier, or several intensifier stages, and a CCD.

metal-oxide silicon (MOS) type of junction used in semiconductor devices.

numerical aperture (NA) the product of the refractive index of the medium between a lens and the object and the cosine of the angle between the outermost ray entering the lens and the optical axis. NA is important because it is proportional to the resolving power of a lens. The 'f number' used by photographers is $1/(2NA)$, i.e. F1 is equivalent to 0.5 N.A.

optical transfer function (OTF) the Fourier transform of the point-spread function.

PARISS trademark of Lightform Inc. for an imaging spectrometer system comprising a curved prism and reflector, giving high spectral resolution in a compact form.

point-spread function (PSF) the mathematical expression that describes the distribution of light from a point source as imaged by a microscope.

quantum efficiency (QE) for example, the number of photoelectrons generated per incident photon (in a detector) or number of photons emitted per photon absorbed (in a fluorochrome).

silicon-intensified target (SIT) combination of an image intensifier, a silicon plate or target and a conventional tube camera, providing low-light-level imaging.

signal-to noise-ratio (SNR) signal level divided by noise level in a detector.

total internal reflectance fluorescence (TIRF) fluorescence imaging in which fluorescence is excited by the evanescent wave spreading into a medium of low refractive index (e.g. saline) from a boundary with a high-index medium (e.g. silica), as a consequence of total internal reflection of an intense beam of exciting radiation at the boundary between the two media.

Index

(references to figures in bold)